SpringerBriefs in Applied Sciences and Technology

PoliMI SpringerBriefs

AF800674

For further volumes:
http://www.springer.com/series/11159
http://www.polimi.it

Fabrizio Ceschin

Sustainable Product-Service Systems

Between Strategic Design and Transition Studies

POLITECNICO DI MILANO

Springer

Fabrizio Ceschin
School of Engineering and Design
Brunel University
Uxbridge, Middlesex
UK

ISSN 2282-2577 ISSN 2282-2585 (electronic)
ISBN 978-3-319-03794-3 ISBN 978-3-319-03795-0 (eBook)
DOI 10.1007/978-3-319-03795-0
Springer Cham Heidelberg New York Dordrecht London

Library of Congress Control Number: 2013955074

Printed on acid-free paper

Springer is part of Springer Science+Business Media (www.springer.com)

Foreword

A New Strategic Design Research Frontier for Sustainability

It is a shared opinion that Product-Service System (PSS) innovations represent a promising approach to sustainability. These offer models have been studied since the end of the 1990s as a win–win one and remarkable advances have been achieved both in knowledge-base and in know-how (methods and tools). However, the application of sustainable PSSs is still very limited! In fact, sustainable PSSs are in most of the cases radical innovations, and so their implementation and diffusion are hindered by several barriers on cultural, corporate and regulative levels.

It has been argued that the diffusion of sustainable PSSs is linked to the attractiveness, acceptance and satisfaction of such alternatives. This should open the debate on the need for an aesthetic of sustainable PSSs as a new aesthetic of interactions and services able to enhance the specific characteristics and the inner qualities of this new generation of artefacts. Others have argued that PSS business models have been mainly researched and implemented as eco-efficiency opportunities in the industrialised context, therefore it should be clarified if and how a PSS approach may couple eco-efficiency with socioethical sustainability. Still others have started to research on the opportunities that a PSS may offer when applied to Distributed Renewable Energy systems, to speed up their diffusion as a key leverage for sustainable development.

This said, it seems to me that that there is a design research stream transversal to the previously mentioned ones, and this research stream is addressed in this book. Sustainable PSSs, being in most cases radical innovations, are often immature when they enter the market, thus, if immediately exposed to market competition of a given context, they have (had in fact) great probability to not survive. For this reason there is a potential role for design not only to propose PSS concepts, but also design transition paths to facilitate and support the experimentation (design the socio-technical experiments), introduction and scaling-up of these concepts. This is very crucial whatever the context we are talking about, or whatever the sustainability dimension we are more concerned of is.

This book is an original and at the same time effective contribution to this latter key issue, exploring and delineating a new strategic role of designers. It is a key contribution for that design (research) arena committed to effectively build up a radically more sustainable and enjoyable society.

Carlo Vezzoli

Preface

The Product-Service System (PSS) concept represents, theoretically and practically, a promising model to steer our production and consumption systems towards sustainability. PSSs are business models based on selling performance (i.e. results) rather than products. They can provide a range of economic and competitive advantages and, if properly designed, can support the dematerialisation of economy and hence provide environmental benefits.

However, it must be stressed that the uptake of this business concept by companies is still very limited. The key problem is that sustainable PSSs can be considered radical innovations. They may require a profound redefinition of the production and consumption modalities and hence their implementation may cope with the current and dominant socio-technical systems. In other words they may challenge existing customer habits, organisational structures and regulative frameworks. For this reason the introduction and scaling-up of such innovations are not completely under the control of a company (or a small network of actors), because changes in the factors that form the boundary conditions (i.e. existing organisations, institutions and networks that share dominant practices, rules and interests), are also required.

Although the concept of sustainable PSS has been discussed in the literature for over a decade, not much attention has been devoted to understand how the process of introduction and scaling-up takes place and how it can be managed and oriented. There is therefore a *knowledge gap regarding the mechanism and factors driving the implementation and diffusion of this kind of innovations and, consequently, there is a lack of strategies, approaches and tools to enable project managers, management consultants and strategic designers in designing and managing this process.* The research presented in this book aims at tackling these issues and in particular at answering the following questions: *How sustainable Product-Service System can be introduced and scaled-up? How this transition process can be designed, managed and oriented?*

The book seeks to answer these challenges by integrating concepts and insights from two research streams: one on *Product-Service System (PSS) design* and the other on *transition studies*. In particular the book, through industrial case studies and an action research project, explores and delineates the role of strategic design in supporting the introduction and scaling-up of sustainable PSS innovations.

A new strategic design role thus emerges, a role in which the ideation and development of sustainable PSS concepts is coupled with the designing of appropriate *transition paths* to gradually incubate, introduce and diffuse these innovations. A key role in these transition paths is given to the implementation of *socio-technical experiments*: protected spaces where radical innovations can be tested, become more mature, and potentially challenge and change dominant socio-technical practices, habits and institutions. The book contributes to clarify how these socio-technical experiments can act as incubators where PSS innovation can start, proliferate and develop. In particular the book focuses on the role of experiments as *Labs* (to test, learn and improve the PSS innovation on multiple dimensions and involving a multiplicity of actors), *Windows* (to raise interest on the innovation project and the related actors, disseminate results, build-up synergies with existing similar projects/initiatives, and attract and enrol new actors), and *Agents of change* (to influence contextual conditions in order to favour and hasten the societal embedding process).

The book outlines and discusses the new design approach and capabilities needed by strategic designers and project managers to design transition paths and socio-technical experiments. Finally, on a more operational point of view, the book presents a practical "how to do it" design process, and associated guidelines, to support practitioners in designing and managing the societal embedding process of sustainable PSS innovations.

The book is structured as follows.

Chapter 1 describes the nature of the problem addressed in this research. It shows that sustainability needs radical innovations, and it argues that Product-Service System (PSS) innovations represent a promising approach to steer the current production and consumption system towards sustainability. The chapter then illustrates the research goals and the research methodology adopted.

Chapter 2 provides an overview of the PSS field. It introduces the PSS concept, describes its potential benefits and the main drivers and barriers to shift towards a PSS-oriented business strategy. The chapter argues that, even if the PSS concept represents a promising economic model to decouple economic value from material and energy consumption, its application is still very limited. Therefore the challenge is not only to conceive sustainable PSS concepts (several methods and tools can in fact be used to support this task), but also to understand which strategies and development pathways are the most appropriate to favour their introduction and scaling-up.

Chapter 3 describes how radical innovations take place, and which are the related dynamics, processes and influencing factors. The chapter then illustrates the concept of socio-technical experiment and in particular its role in triggering transition processes. The chapter also discusses to which extent the concepts and insights from transition studies are valuable for the specific characteristics of sustainable PSS innovations. As a result, the chapter puts forward a conceptual framework to describe the process of implementing and diffusing sustainable PSS innovations, and its main influencing factors. The argumentation is accompanied by the illustration of several case studies.

Chapter 4 investigates the potential contribution that a strategic design approach can make in stimulating and supporting the societal embedding of sustainable PSSs. A new strategic design role emerges. A role in which the ideation and development of sustainable PSS concepts is coupled with the designing of appropriate transition paths to gradually incubate, introduce and diffuse these concepts. Starting from these considerations the chapter outlines and discusses the new design approach and capabilities required by strategic designers/project managers/consultants. The argumentation is accompanied by the discussion of an experimental design experience: the Cape Town Sustainable Mobility project.

Chapter 5 summarises the main findings and lessons learned, highlights and generalises the main contributions, and indicates potential paths for future research.

This book is thought for a broad audience interested in radical innovations towards sustainable production and consumption systems.

Firstly, the book may be of interest to the *academic and research communities* studying Product-Service Systems and (more in general) sustainable innovations and strategies, from both the design and managerial perspectives. The research provides in fact new theoretical knowledge and practice-oriented insights into how to understand, design and manage the societal embedding process of sustainable PSSs. This book is also aimed at the academic and research community in the field of transition studies: in fact it presents approaches and tools that could be adapted and used more in general to manage the introduction and scaling-up of radical innovations.

Secondly, the research may be useful for *practitioners* (strategic designers, project managers and consultants) who want to acquire the basic knowledge and the practical 'how to do it' competence to support and guide a company (or a small network of actors) in managing and enhancing the societal embedding process of eco-efficient and sustainable PSS innovations.

For the same reasons, this book is also aimed at *companies* that are shifting from product-based towards PSS-oriented business strategies, and want to acquire knowledge and competences about how to increase the chances to successfully introduce and scale-up PSS concepts.

Finally, the study may be of interest to *graduated students* (in the fields of design and management) dealing with PSSs and sustainable innovations.

Fabrizio Ceschin

Acknowledgments

I would like to express my gratitude to the people who accompanied me during this research journey.

First of all, I would like to thank Carlo Vezzoli, for his insightful guidance and his invaluable support and encouragement throughout this research.

During this research I spent some months in the department of *Industrial Engineering and Innovation Sciences* of the *Eindhoven University of Technology*, where I had the opportunity to discuss my work and gain valuable insights. For this reason I would like to express my gratitude to Rob Raven for sharing his ideas and for offering his constructive comments and precious advice.

I am very grateful to the people who were (and still are) engaged in the *Cape Town Sustainable Mobility* project: it was intellectually stimulating and a real pleasure to collaborate with Mugendi M'Rithaa, Shona McDonald, Guillaume Du Toit, Andrew Wheeldon, Fagodien Campher and Filippo Sala. In particular a special thank you goes to Emanuela Delfino and Silvia Remotti for all the enthusiam, energy and effort they put on the project.

I would like to thank the *Learning Network on Sustainability (LeNS)* crew that stimulated this research and provided critical comments at different stages. In particular I am grateful to Jan Carel Diehl for his useful insights into how to structure the research approach and methodology at the first stages of this journey.

I wish to express my appreciation to Anna Meroni, Ezio Manzini and Luisa Collina for sharing their experience and insightful thoughts.

A big thank you goes to the researchers, designers and experts who provided useful critics and insights during the research process: Agnese Brasca, Maurizio Catulli, Chiara Del Gaudio, Emanuela Delfino, Idil Gaziulusoy, Dries Hegger, Boukje Huijben, Hussain Indorewala, Yong Se Kim, Benny Leong, Tatu Martilla, Daniel Metcalfe, Mugendi M'Rithaa, Ravi Mokashi, Mariano Ramirez, Mauro Rego, Silvia Remotti, Arnold Tukker, Philip Vergragt and Liu Xin.

Finally I would like to thank Krystal for her love, for taking care of me and for being there during the whole research journey.

Contents

Chapter 1
Transition Towards Sustainability: The Need of Radical Innovations

Abstract This chapter describes the nature of the problem addressed in this research. It shows that sustainability problems needs radical innovations, and it argues that Product-Service System (PSS) innovations represent a promising approach to steer the current structure of production and consumption towards sustainability. The chapter then illustrates the research goals: to understand how sustainable PSS innovations can be introduced and scaled-up, and how strategic design may contribute in triggering and supporting this process. The chapter then describes the research methodology adopted to answer the research questions.

Keywords Sustainable development · Radical innovation · System innovation · Regime shift · Design for sustainability · Transition studies · Product-service system

1.1 The Sustainability Challenge

Our planet has existed for more than 4.5 billions of years. If 100 represents the whole lifetime of Earth, human being was born only on the 99th year, 4–5 millions of years ago. Despite this relatively short period of time humans' capabilities of adapting the natural environment in which they live has determined profound modifications on Earth's biosphere. The degree of these changes grew-up exponentially starting from the first industrial revolution. Since then the pressure of human being on natural environment has been so devastating that the Nobel Prize in chemistry Crutzen and his colleague Stoermer suggested the industrial revolution as the beginning of a new geological period: the "*Anthropocene*" (Crutzen and Stoermer 2000).

After decades in which natural resources were considered inexhaustible and the resilience capacity of the Earth was not an issue, we are now fully aware of the effects that our actions have produced, and still produce, on ecosystems. The discovery of the finite nature of resources led to realise that the dominant

F. Ceschin, *Sustainable Product-Service Systems*, PoliMI SpringerBriefs,
DOI: 10.1007/978-3-319-03795-0_1, © The Author(s) 2014

socio-economic development models in industrialised countries, based on producing and selling goods, cannot be sustained by our planet. A fundamental contribution to this understanding came in 1972 from the report commissioned by the Club of Rome to the Systems Dynamics group of the Massachusetts Institute of Technology. The results of the research, named *The limits to Growth* (Meadows et al. 1972), showed, through simulation models, the effects of the system of production and consumption on nature; it was the first scientific forecast of a possible global ecosystem collapse. Overpopulation, increasing resources use and pollution, and ever increasing consumption were identified as the three main trends aggravating the problem (ibid.). In the introduction of the report the executive committee of the Club of Rome stated that the research "*indicates that humanity cannot continue to proliferate at accelerate pace, considering the material growth as the main aim, without facing the natural limit of the process, in front of which humanity can choose to take new paths to control the future, or to accept the most cruel unavoidable consequences of an uncontrolled growth*". Still today we face the dangers of the environmental limits and the irreversibility of harmful effects. As stated in the updated report published in 2004 (Meadows et al. 2006): "*The result is that today we are more pessimistic about the future than we were in 1972. […]. We must change many things if we don't want that the overcoming of the limits will lead to a collapse in the 21st century*".[1] Translated in other terms this means that we must be able to move from a society in which wellbeing is measured by the production and consumption of goods, to one in which people live better consuming (much) less (Vezzoli and Manzini 2008).

What indicator can we use to show we cannot go on with our current development model? We can choose any one of the following: carbon in the atmosphere,[2] energy consumption,[3] resources availability,[4] lost topsoil, accessibility to drinking water, lost biodiversity[5]… We will only say that the Ecological Footprint is increasing and it is beyond the Earth biocapacity: 1.5 years are needed to regenerate the renewable resources used in 2007 (WWF 2010). The size of the global ecological footprint in 2007 is doubled compared to 1966 and, under a *business-as-usual* scenario, the outlook is that by 2030 humanity will need the capacity of 2 Earths to absorb CO_2 waste and keep up with natural resource consumption (ibid.). Also, it is not only a matter of resources consumption but also a matter of resources distribution: in fact 20 % of the world's richest countries consume 80 % of the resources. If everyone in the world lived like an average

---

[1] Translated into English from the Italian version of the book.

[2] In 2007 the CO_2 level reached the 400 ppm, increasing by 40 ppm in ten year. This change of magnitude over only a decade has not been seen since the most recent ice age ended around 10,000 years ago (May 2007).

[3] Energy consumption soared by 68 % from 1990 to 2010 (Enerdata 2011).

[4] Some critical resources will run out within 10 years (REKTN 2008).

[5] The Living Planet Index reflects changes in the health of the planet's ecosystems by tracking trends in nearly 8,000 populations of vertebrate species. The last Living Planet Report (WWF 2010) shows that vertebrate species populations declined of around 30 % from 1970 to 2007.

resident of the United States or the United Arab Emirates, then a biocapacity equivalent to more than 4.5 Earths would be required to keep up with humanity's consumption and CO_2 emissions (ibid.). On the other hand, if everyone lived like the average resident of India, humanity would be using less than half the planet's biocapacity (ibid.).

In short, as stated in the: Millennium Ecosystem Assessment (2005) "*Many people have benefited over the last century from the conversion of natural ecosystems to human-dominated ecosystems and from the exploitation of biodiversity. At the same time, however, these gains have been achieved at growing costs in the form of losses in biodiversity, degradation of many ecosystem services, and the exacerbation of poverty for other groups of people.*"

The question at this point is: what is the scale of change required to steer our production and consumption systems towards sustainability?

In the second half of the 1990s a series of studies and analyses led to a clearer understanding of the change necessary to achieve a society that is effectively and globally sustainable (Factor 10 Club 1994; Vergragt and van Grootveld 1994; Schmidt-Bleek 1996). Taking into account an increase in the demand for wellbeing in currently emerging and low-income contexts, these studies have brought out an astonishing result: conditions for sustainability are achievable only by increasing the efficiency of the production and consumption systems in mature industrialised countries by at least ten times. These estimates are approximate. However, they are valid to indicate the scale of change that should take place. We can only consider sustainable those socio-technical systems capable to drastically reduce the consumption of environmental resources (compared to the average consumption levels in mature industrialised contexts), and equally distribute them. In other terms a profound, radical transformation in our development model is necessary.

In addition, we have to be aware that it is not only a matter of scale of change but also a matter of pace of change. In fact, as underlined by Evans et al. (2008), the window of opportunity is rapidly closing because the Earth is coming under stress more quickly than predicted. We have therefore to urgently move towards a system capable to satisfy people needs while operating within planet limits.

The concept of sustainability, based on economic, environmental and social inter- and intra-generational equity (WCED 1987), is today widely shared. However, the same cannot be said about the paths to achieve it. In fact the complexity of the issue and the diversity of views make it difficult to define widely acceptable strategies. Thus, the question is: how can we tackle the sustainability challenge?

1.2 Evolution of Approaches to Sustainability

During the last decades the reaction of humankind to sustainability problems has moved from end-of-pipe approaches to actions increasingly aimed at prevention. Four generations of approaches can be identified (Simons et al. 2001).

The *first generation* of approaches, in the late 1960s, took the form of pollution control and end-of-pipe solutions. These approaches try to control the pollution after it has occurred. They focus on remediating the environmental effects caused by human activities (e.g. clean up a lake polluted by an industrial plant), without questioning on the causes of the problems.

Since the mid 1980s environmental approaches has shifted the attention from the control, management and treatment of pollutants and waste at the end of the process, to the prevention of pollution and waste generation. These kinds of interventions, characterised by a shift from reactive to proactive solutions, represent the *second generation* of approaches. They are defined *cleaner production strategies* (UNEP 1994), and aim to prevent environmental effects by improving (industrial) processes (e.g. use clean technologies to avoid the industrial plant to pollute the lake). These strategies include technological innovations and improvement of organisation in order to e.g. conserving raw materials and energy, eliminating toxic raw materials, reducing the quantity and toxicity of all emissions and wastes before they leave the process, on site recycling, etc. (USEPA 1992). The result of these strategies is an increased eco-efficiency of the production processes.

However, since products determine environmental effects along their whole life cycles, reducing the impact solely on the production phase can only lead to limited benefits. For this reason, in the early 1990s the attention slowly focused not only on production processes but also on products (and their life cycles). These new approaches, focused on reducing the environmental impact of products' life cycle, constitute the *third generation* of approaches. Several labels have been used to identify them. The most commonly used are *eco-design* (Brezet and van Hemel 1997), *Design for Environment (DfE)*, and *Life Cycle Design (LCD)* (Manzini and Vezzoli 1998). These approaches built upon new insights about the environmental effects linked to the production, use and disposal of a product, and on new methods of assessing these impacts (i.e. *Life Cycle Assessment, LCA*). The approaches and tools to design low environmental impact products have rapidly evolved since the early 1990s, and allow companies and designers to effectively integrate environmental design requirements into the product development process.

The strategies included in the first three generations of approaches have one common characteristic: they do not modify the structures of production and consumption but they only optimise them. For this reason, although they are fundamental and necessary, they are not alone sufficient to obtain the radical shift required to achieve sustainability conditions (the previously mentioned reduction of 90 % of resources consumption). In fact, even if these innovations can improve the environmental performance of products, it is also true that these improvements are often negatively counterbalanced by an increase in consumption levels (Schmidt-Bleek 1996; Brookes 2000; Binswanger 2001). In other words, as underlined by Mont (2004), these approaches refer in general to the process but not to the quantity of output. Several examples can point up this problem. For instance, the environmental gain achieved through the improvement of car efficiency in the last 15 years (10 %) has been more than offset by the increase in the number of cars and by the correlated increase (30 %) in the overall amount of km covered (EEA 2008). A similar example can be found in the energy-efficient household appliances, which become more and more efficient per unit of volume; on the other hand the increase in the appliances volume and in the number of sold appliances led to increases in aggregate energy consumptions (Mont 2004).

In addition, another problem is related to the fact that in the traditional production and consumption model, based on the production and sale of products, producers are usually not economically interested in optimising/reducing the consumption of resources along the product's life cycle (UNEP 2002). For instance car producers are not economically incentivised in extending cars lifespan as much as possible (on the contrary they are interested in accelerating the substitution of spare parts and the replacement of the car itself in order to increase profits).

In short, as underlined by Hawken et al. (1999): "*Without a fundamental rethink of the structure and the reward system of commerce, narrowly focused eco-efficiency could be a disaster for the environment overwhelming resource savings with even larger growth in the production of the wrong products, produced by the wrong processes, at the wrong scale and delivered using the wrong business models*".

Ehrenfeld (2008) argues that the strategies included in the first three generations of approaches constitute symptomatic solutions which do not go to the root of the problem. The more we follow these approaches and the more we have the unjustified expectations that they work. In reality, sooner or later the problems to which they have been applied will either re-emerge or worsen (side effects) (ibid.). Using his words, "*reducing unsustainability will not create sustainability*".

Because of these reasons, if we want to effectively tackle sustainability, there is a need to move from a focus on product improvements only, towards a wider approach focused on producing structural changes in the way production and consumption systems are organised. These strategies, which focus at a system innovation level, represent the *fourth generation* of approaches. Their assumption is that sustainability issues must be tackled focusing on the root of the problem and not on its symptoms.

This means that a profound radical transformation of our development model is necessary. As underlined by Manzini (1999), under discussion there are not only production processes and artefacts (products and services, infrastructure and all various form of anthropological settlements) but also patterns of consumption and access to goods and services. There is a need to develop new potential ways of satisfying needs and desires, and in general the social demand of wellbeing. It is difficult to foresee how this may happen, but there will probably be a discontinuity that will affect all facets of the system (Vezzoli and Manzini 2008). Given the nature and the dimension of this change, the transition towards sustainability represents a wide-reaching social learning process in which system innovations are needed. This implies changes at different levels: at the level of the products, services and production systems, and equally at the level of social and institutional arrangements, such as mechanisms of coordination (regulation, governance) and patterns of interaction at the supplier and the user side of innovation (Weber et al. 2006). Thus, significant changes in the way we think are required (Evans et al. 2008), because the same mind-set that led us to the current environmental, social and economic problems cannot be used to solve them. As underlined by Ehrenfeld (2008), it is a matter of changing the culture and breaking down the routine behaviours that are daily reproduced by individuals, groups, business communities, policy actors and society at large.

The question at this point is: in which way we can re-shape our production and consumption system to steer it towards sustainability?

1.3 Functional Economy: A Promising Model

A concept that theoretically and practically represents a promising model to steer our production and consumption systems towards sustainability is the *functional economy* one. Using Stahel's words: "*Functional economy is an economy that optimises the use (or function) of goods and services and thus the management of existing wealth (goods, knowledge, and nature). The economic objective of the functional economy is to create the highest possible use value for the longest possible time while consuming as few material resources and energy as possible. The functional economy is therefore more sustainable, or dematerialised, than the present economy, which is focused on production as its principal means to create wealth and material flow*" (Stahel 1997).

In the perspective suggested by Stahel we should move from an *industrial economy*, in which the central value is based on the exchange of products to be consumed and in which the growth is strongly linked to resource consumption, to a *functional economy*, in which products and technology are mere modes of providing functions (Mont 2002). Functional economy is oriented to satisfy consumers through the delivery of functions instead of products (e.g. mobility instead of cars, thermal comfort instead of heating systems), and this can potentially bring about a reduction in the current levels of resources consumption, without minimizing consumer level of satisfaction.

It is a model that changes the rewards system of the commerce. In a functional economy in fact, producers and providers are paid per unit of function delivered and not per unit of product sold. In other words the objective of the sale is to deliver a performance to provide customer satisfaction. Less material and economic resources are used to deliver that satisfaction and higher will be the profits for the provider. Thus, the functional economy model can provide an economic incentive to reduce and optimise the amount of material and energy consumption. In other terms the satisfaction of customer needs and the economic health can potentially be decoupled from the material and energy consumption.

1.4 The Product-Service System Concept and its Sustainability Potential

Within the functional economy framework it is considered promising to look at the concept of Product-Service System (PSS) innovation, understood as "*the result of an innovative strategy that shifts the centre of business from the design and sale of (physical) products alone, to the offer of product and service systems that are together able to satisfy a particular demand*" (UNEP 2002). PSS is not a new economic concept: several examples of PSSs have been implemented in the last decades by various companies.[6] From a sustainability point of view the key point to be underlined is that, if properly conceived, a PSS can decouple economic value from material and energy consumption[7] (White et al. 1999; Stahel et al. 2000; Heiskanen and Jalas 2000; Wong 2001; Zaring et al. 2001; UNEP 2002). PSSs can in fact offer an economic and competitive incentive for firms involved to continuously foster improvements in resource productivity. In addition to potentially decouple value creation from resources consumption, a PSS-oriented business strategy can also provide market opportunities and an improved strategic positioning for the companies involved (Goedkoop et al. 1999; Wise and Baumgartner 1999; Manzini et al. 2001; UNEP 2002; Mont 2002). PSS solutions can also provide a wide range of socio-ethical benefits (Tukker et al. 2006; Penin 2006; Vezzoli 2007). Moreover this kind of business models is promising for both industrialised contexts and emerging and low-income ones (UNEP 2002; Penin 2006).

It must be stressed out that the *functional economy* and the PSS concepts present some points of contact with other sustainability concepts:

- the *natural capitalism* concept, which sees a sustainable industrial system as being based on four main pillars: "*radically increased resource productivity, redesigning industry based on biological models with closed loop and zero waste, shifting from the sale of goods to the provision of services, and reinvesting in natural capital*" (Hawken et al. 1999);
- the *cradle to cradle* concept, a specific form of industrial ecology whereby all materials used in the system are separated into biological nutrients and technical nutrients; the first ones can be decomposed and allowed to re-enter the natural system, while the second ones should be kept within the industrial system and used multiple times (McDonough and Braungart 2002);
- and the *systemic design* concept (Bistagnino 2011; Pauli 2010) which sees production and consumption systems as open systems where the output (waste) of a system becomes input for another one. The aim is to create self-standing systems capable to sustain themselves thanks to the re-use and optimisation of

---

[6] For example see the cases collected by Goedkoop et al. (1999), UNEP (2002), Mont 2004), Vezzoli (2007).

[7] It must be stressed out that not all PSSs are sustainable. They have a potential to contribute to sustainability only if carefully designed. This will be better explained in Chap. 2.

material and energy flows coming out and in the actors belonging to the systems. Special attention is moreover given to the local context and its human, cultural and material resources.

Finally it has to be stressed out that although the PSS concept seems to be a valuable and promising concept to tackle sustainability issues, it does not represent a silver bullet. Of course it can provide a wide range of economic, environmental and socio-ethical benefits, but it cannot alone solve all the problems. The author's opinion is that synergies have to be built with other promising and interwoven concepts. In particular the *social innovation*[8] and *distributed economies*[9] ones.

1.5 Research Scope and Questions

A wide number of research projects in the field of PSS and sustainability have been recently supported by EU funding, and several design methods and tools have been developed in the last years to orient and support the development of sustainable PSS. However, despite all the knowledge accumulated, it has to be underlined that the uptake of this business concept by companies is still very limited. The reason is that sustainable PSSs can be considered, in most of the cases, **radical innovations**, because they challenge existing customer habits, organisational structures and regulative framework*s*. In other words, their introduction and scaling up require breaking down the routine behaviour that is daily produced by individuals, groups, business communities, policy actors and society at large.

For this reason the introduction and scaling-up of such innovations are not completely under the control of a company (or a small network of actors), because changes in the factors that form the boundary conditions (i.e. existing organisations, institutions and networks that share dominant practices, rules and interests), are as well required. In other words sustainable PSS innovations require changes in the socio-technical context, and hence cannot be realised by only asking a single

---

[8] Social innovation is a process of change whereby new solutions emerge from a variety of actors directly involved in the problem to be solved: communities, grass roots technicians and entrepreneurs, local institutions and civic society organizations. Some of the solutions emerged (some examples are: self-managed services for the care of children and the elderly; new forms of exchange and mutual help; alternative mobility systems; networks linking consumers directly with producers, etc.) present interesting potentialities in generating and diffusing new and more sustainable ways of living. This kind of innovation has always existed. But now there are good reasons to say that its role is expanding and will expand in the near future. For details see Manzini (2005), Meroni (2007), and Jégou and Manzini (2008).

[9] Distributed economies are activities organized in the form of small scale, flexible units that c synergically connected with each other (e.g. distributed energy systems). They are potentially able to make optimum use of local resources, both physical and social ones, and therefore to reduce their environmental impact and increase social inclusion and democratic participation. For details see Johansson et al. (2005).

company to change its business model. Therefore the challenge is not only to conceive sustainable PSS concepts (several methods and tools can in fact be used to support this task), but also to understand which strategies and development pathways are the most appropriate to favour their introduction and scaling up. Although the concept of sustainable PSS has been discussed in the literature for over a decade, not much attention has been devoted to understand how the process of introduction and scaling-up takes place and how it can be managed and oriented. There is in fact *a knowledge gap regarding the dynamics, mechanism and factors driving the implementation and diffusion of this kind of innovations* and, consequently, there is a lack of strategies, approaches and tools to enable *strategic designers, project managers* and *management consultants* in designing, managing and orienting this process. This study focuses on this unexplored research area.

In order to address the mentioned gaps, the following *research questions* are formulated:

1. How do sustainable Product-Service System innovations take place?
    (a) What are the dynamics and factors that facilitate and hinder the process of introduction and scaling-up?
    (b) Is it possible to manage and orient this process? And if yes, how?
2. Can strategic design have a role in supporting and orienting this process?
    (a) If yes, what kind of knowledge base and capabilities are needed by a strategic designer?
    (b) From an operational point of view, what is the design approach, methods and tools that can be used in practice?

Recent advancements in the *transition studies* field[10] have provided insights into how to understand, influence and orient the adoption of system innovations. According to these theories, radical innovations are often immature when they enter the market because they cope with a dominant socio-technical context (and its established and stable rules and networks of actors), and if immediately exposed to market competition, they have great probability to not survive. For this reason the introduction of radical innovations requires the creation of partially protected socio-technical experiments where the innovation can be tested, incubated and brought to mature (Kemp et al. 1998; Hoogma et al. 2002; Brown et al. 2003). The purposive creation of experiments that are protected from commercial market competition can help to overcome the innovation inertia enforced by the prevailing cultural, organizational and regulatory rules, which are referred to as an incumbent dominant regime. A pathway of socio-technical experiments can be used as a strategic arena for learning, shaping future expectations and establishing

[10] In particular the contributions from Strategic Niche Management (SNM) and Transition Management (TM) approaches.

new social networks in order to gain momentum for diffusion and challenge and change socio-technical regimes (Raven 2005).

The assumption upon which this research is built is that these theoretical concepts can provide insights on how the introduction and diffusion of eco-efficient and sustainable PSS innovations take place and how this process can be oriented and stimulated. In other words they can be used to understand the dynamics and factors that drive and orient this kind of radical innovations. Thus, this research proposes the adoption and adaptation of concepts and insights from transition studies into PSS design. The hypothesis is that there are some potential synergies between *PSS design for sustainability* and *transition studies*, and that through these synergies it may be possible to build up a proper set of knowledge and know-how capable to equip and support strategic designers in designing, managing and orienting the process of introduction and diffusion of this kind of innovations.

1.6 Research Methodology

When aiming at addressing the above research questions, the main challenge is related to the nature of sustainable PSS innovations. They are in fact complex phenomena: they represent change processes over time and they are multi-factor and multi-actor. Thus, they are difficult to be measured quantitatively and cannot be reproduced and studied in a lab setting. Also, the process of introduction and scaling-up of sustainable PSSs might require several years. Therefore the implementation and testing of design approaches/strategies cannot be easily studied in a real time perspective. Because of these reasons, qualitative research was selected to tackle these questions. In fact qualitative research is suited to explore complex phenomena that are difficult to measure quantitatively, and to generate insights necessary for a comprehensive understanding of a problem (Patton 2002; Crabtree and Miller 1999). In particular, in order to tackle this challenge, the adopted methodology is based on three main stages (Fig. 1.1)[11]:

- *Stage* 1: *Case studies research on sustainable PSS innovations*. The aim of this stage was to understand how the process of introduction and diffusion of sustainable PSS innovations take place (which are the dynamics and mechanism and the associated influencing factors), and how it can be successfully managed. Thus, this stage sought to provide an answer to the first set of research questions. A **literature review** on transition studies was conducted to identify the potentially valuable insights for understanding and managing the implementation and diffusion of sustainable PSSs. These insights were used to develop a **conceptual framework** aimed at describing how sustainable PSS innovations take place and at providing an overview of the factors which influence the process (and how these factors are interrelated). Then, a **case studies research** was used to

[11] For an in depth discussion of the research methodology see Ceschin (2012), pp. 53–70.

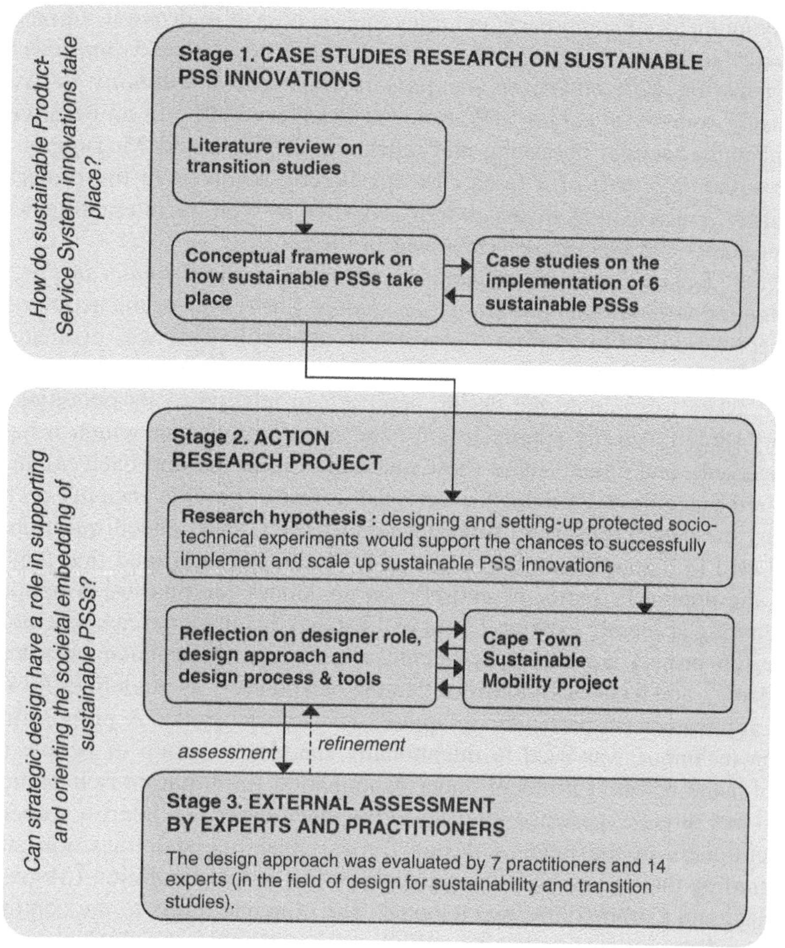

Fig. 1.1 Research methodology: overview of stages and main activities

validate, adjust and refine the conceptual framework. In particular, the case studies research analysed the process of implementation and scaling-up of six sustainable PSSs.

- *Stage* 2: *Action research project aimed at designing, introducing and diffusing a PSS innovation*. Building upon the results of the previous stage, the aim of this stage was to develop a design approach (and related method and tools) to enable strategic designers in supporting and facilitating the process of implementation and diffusion of sustainable PSSs. An **action research project**, aimed at designing and introducing a sustainable mobility system in the suburban areas of Cape Town, was used to test and reflect on the design approach, as well as to develop insights on how to refine and make it more applicable to practice. This was an iterative process in which researchers were continuously involved in applying the design approach and reflecting on how to improve it. In fact action research seeks to "*bring together action and reflection, theory and practice, in participation with others, in the pursuit of practical solutions to issues of pressing concern to people*" (Reason and Bradbury 2001), in an iterative cycle of planning, acting, observing and reflecting (Kemmis and McTaggart 1988). The author was part of a research team directly involved in the project management, participating in the design activities as well as interacting with the other actors and practitioners involved in the project.

- *Stage* 3: *External assessment made by experts and practitioners in the field of design for sustainability and transition studies.* The results of the action research project were used to propose a new design approach. This was evaluated by 7 practitioners and 14 experts through a questionnaire. In particular participants were asked to evaluate the design approach in relation to its potential practicality (how much the approach is usable in the settings for which it has been conceived), and effectiveness (how much the use of the approach might led to desired outcomes). Participants were also asked to provide insights on how to refine the design approach. Both open-ended and closed-ended questions were included in the questionnaire. 6 out of 7 practitioners affirmed that they could use the approach (partly or entirely) as guidelines for on-the-job application. Regarding academic experts, 13 out of 14 stated that the approach and the action research project are useful as reference material (in particular as a teaching resource), and 9 out of 14 affirmed that they could use it as guidelines for on-the-job application (in particular in applied research projects). A purposeful sampling technique was used to intentionally sample the group of people to best inform the research problem under examination (information-rich participants who had direct experience relevant to the phenomenon of interest: experts and practitioners in the field of design for sustainability and transition studies). Regarding the sample size, the principle of theoretical saturation (Morse 1995; Strauss and Corbin 1998) was adopted. The principle refers to the continuation of sampling and data collection until no new conceptual insights are generated. In particular, a sample size of 14 academic experts and practitioners was initially used. Then 7 additional participants were involved. Based on the information provided by the second group of participants and the scarcity of new information emerged from the last 3 questionnaire (comments and suggestions for improvement were similar to the one provided by the other participants), sampling was completed with 21 participants.

In sum, the process to answer to the research questions was not linear but rather iterative and interactive. *Iterative*, because the design approach was implemented in a practical design experience and continuously developed, adjusted and refined during the whole design process. *Interactive*, because the process was characterised by a continuous collaboration among researchers, practitioners and experts (which continuously brought input on how to improve the design approach and tools).

References

Binswanger M (2001) Technological progress and sustainable development: what about the rebound effect? Ecol Econ 36:119–132

Bistagnino L (2011) Systemic design. Slow Food Editore, Bra

Brezet H, van Hemel C (1997) Ecodesign a promising approach to sustainable production and consumption. UNEP, Paris

Brookes L (2000) Energy efficiency fallacies revisited. Energy Policy 28:355–366

Brown HS, Vergragt PJ, Green K, Berchicci L (2003) Learning for sustainability transition through bounded socio-technical experiments in personal mobility. Technol Anal Strateg Manag 13:298–315

Crabtree B, Miller W (eds) (1999) Doing qualitative research. Sage, Calif

Ceschin F (2012) The introduction and scaling up of sustainable product-service systems. A new role for strategic design for sustainability. PhD dissertation, Politecnico di Milano

Crutzen PJ, Stoermer EF (2000) The "Anthropocene". IGBP Newsletter 41:17–18

Ehrenfeld JH (2008) Sustainability by design. A subversive strategy for transforming our consumer culture. Yale University Press, New Haven and London

Enerdata (2011) Global energy statistical yearbook 2011. Available online http://yearbook.enerdata.net/. Accessed 12 Aug 2011

Evans S, Bergendahl MN, Gregory M, Ryan C (2008) Towards a sustainable industrial system. With recommendations for educations, research, industry and policy. University of Cambridge, Cambridge

Factor 10 Club (1994) Declaration of the factor 10 club. Available online http://www.factor10-institute.org/files/factor_10_club/F10Club-Statement_1994_e.doc. Accessed 12 Aug 2011

Goedkoop M, van Halen C, te Riele H, Rommes P (1999) Product services systems, ecological and economic basics, report 1999/36. VROM, The Hague

Hawken P, Lovins AB, Lovins LH (1999) Natural capitalism: the next industrial revolution. Earthscan, London

Heiskanen E, Jalas M (2000) Dematerialisation through services: A review and evaluation of the debate. Ministry of the Environment, Environmental Protection Department, Helsinki

Hoogma R, Kemp R, Schot J, Truffer B (2002) Experimenting for sustainable transport: the approach of strategic niche management. Spon Press, London

Jégou F, Manzini E (eds) (2008) Collaborative services. Social innovation and design for sustainability. Edizioni Polidesign, Milan

Johansson A, Kisch P, Mirata M (2005) Distributed economies. A new engine for innovation. J Cleaner Prod 13:971–979

Kemmis S, McTaggart R (1988) The action research planner. Deakin University Press, Geelong

Kemp R, Schot J, Hoogma R (1998) Regime shifts to sustainability through processes of niche formation: the approach of strategic niche management. Technol Anal Strateg Manag 10:175–195

Manzini E (1999) Sustainable solutions 2020—systems. In: Proceedings of the 4th international conference towards sustainable product design, Borschette Conference Centre, Brussels, 12–13 Jul 1999

Manzini E (2005) Enabling solutions for creative communities: social innovation and design for sustainability. Des Matters 10:64–68

Manzini E, Vezzoli C (1998) Lo Sviluppo di Prodotti Sostenibili. I Requisiti Ambientali dei Prodotti Industriali. Maggioli Editore, Rimini

Manzini E, Vezzoli C, Clark G (2001) Product service systems: using an existing concept as a new approach to sustainability. J Des Res 1:12–18

May R (2007) Hard choices. In: Edwards M (ed) Hard rain. Our headlong collision with nature. Still Pictures Moving Words, London

McDonough W, Braungart M (2002) Cradle to cradle. Remaking the way we make things. North Point Press, New York

Meadows DH, Meadows D, Randers J, Behrens WWI (1972) The limits to growth. Universe Books, New York
Meadows DH, Meadows D, Randers J (2006) I nuovi limiti dello sviluppo. Mondadori, Milano
Meroni A (ed) (2007) Creative communities. People inventing sustainable ways of living. Edizioni Polidesign, Milan
Millennium Ecosystem Assessment (2005) Ecosystems and human well-being: wetlands and water synthesis. World Resources Institute, Washington
Mont O (2002) Clarifying the concept of product-service system. J Cleaner Prod 10:237–245
Mont O (2004) Product-service systems: panacea or myth? PhD dissertation, Lund University, Sweden
Morse J (1995) The significance of saturation. Qual Health Res 5:147–149
Patton MQ (2002) Qualitative research and evaluation methods, 3rd edn. Sage, Calif
Pauli G (2010) The blue economy. 10 years, 100 innovations, 100 million jobs. Report to the Club of Rome, Paradigm Publications, New Mexico
Penin L (2006) Strategic design for social sustainability in emerging contexts. PhD thesis, Poliecnico di Milano
Raven RPJM (2005) Strategic niche management for biomass. Technische Universiteit Eindhoven, PhD dissertation
Reason PE, Bradbury H (eds) (2001) Handbook of action research: participative inquiry and practice. Sage, London
Resource Efficiency Knowledge Transfer Network (2008) Material security technology strategy board. Ensuring resource availability for the UK economy. C-Tech Innovation Ltd., Chester
Schmidt-Bleek F (1996) MIPS book or the fossil makers—factor 10 and more. Boston, Basel, Berlin
Simons L, Slob A, Holswilder H, Tukker A (2001) The fourth generation: New strategies call for new eco-indicators. Environ Qual Manage 11:51–61
Stahel WR (1997) The functional economy: Cultural and organizational change. In: Richards DJ (ed) The industrial green game. National Academy Press, Washington
Stahel WR et al (2000) From manufacturing industry to a service economy, from selling products to selling the performance of products. Product-Life Institute, Geneva
Strauss A, Corbin J (1998) Basics of qualitative research: Techniques and procedures for developing grounded theory. Sage, Newbury Park
Tukker A, Tischner U, Verkuijl M (2006) Product-services and Sustainability. In: Tukker A, Tischner U (eds) New business for Old Europe. Product services, sustainability and competitiveness. Greenleaf Publishers, Sheffield
United Nations Environmental Programme (UNEP) (1994) Government strategies and policies for cleaner production. UNEP, Paris
United Nations Environmental Programme (UNEP) (2002) Product-service systems and sustainability. Opportunities for sustainable solutions. UNEP, Paris
United States Environmental Protection Agency (USEPA) (1992) Facility pollution prevention planning guide. USEPA, Cincinnati
Vergragt PJ, van Grootveld G (1994) Sustainable technology development in the Netherlands. The first phase of the Dutch STD program. J Cleaner Prod 2:133–139
Vezzoli C (2007) System design for sustainability. Theory, methods and tools for a sustainable "satisfaction-system" design. Maggioli Editore, Rimini
Vezzoli C, Manzini E (2008) Review: design for sustainable consumption and production systems. In: Tukker A, Charter M, Vezzoli C, Stø E, Andersen MM (eds) System innovation for sustainability 1. Perspectives on radical changes to sustainable consumption and production. Greenleaf Publishing, Sheffield
White AL, Stoughton M, Feng L (1999) Servicizing: the quiet transition to extended product responsibility. Tellus Institute, Boston
Weber M, Kubeczko K, Rohracher H (2006) System innovations in innovation systems. Experiences with adaptive foresight in Austria. In: Proceedings of the workshop perspectives on radical changes to sustainable consumption and production, Copenhagen, 20 Apr 2006

Wise R, Baumgartner P (1999) Go downstream e the new profit imperative in manufacturing. Harvard Bus Rev 77:133–141

Wong M (2001) Industrial sustainability (IS) and product service systems (PSS): A strategic decision support tool for consumer goods firms. University of Cambridge, Cambridge

World Commission on Environment and Development (WCED) (1987) Our common future. Oxford University Press, Oxford

World Wide Fund for Nature (WWF) (2010) Living planet report 2010. Biodiversity, biocapacity and development, World Wide Fund for Nature, Gland

Zaring O, Bartolomeo M, Eder P, Hopkinson P, Groenewegen P, James P, de Jong P, Nijhuis L, Scholl G, Slob A, Örninge M (2001) Creating eco-efficient producer services. Gothenburg Research Institute, Gothenburg

Chapter 2
Product-Service System Innovation: A Promising Approach to Sustainability

Abstract This chapter provides an overview of the Product-Service Systems (PSS) field. It introduces the PSS concept, describes its potential benefits (in terms of economic, environmental and socio-ethical sustainability), and its main drivers and barriers. The PSS concept represents a promising economic model to decouple economic value from material and energy consumption. However, the adoption of sustainable PSSs by industry is still very limited. The reason is that this kind of innovations can be considered, in most of the cases, radical ones, because they challenge existing customer habits, organisational structures and regulative frameworks. Therefore, the challenge is not only to conceive sustainable PSS concepts (several methods and tools can in fact be used to support this task), but also to understand which strategies and development pathways are the most appropriate to favour their introduction and scaling up.

Keywords Product-service system · Sustainable development · System innovation · Design for sustainability · Functional economy · PSS benefits · PSS barriers · PSS drivers

2.1 Product-Service System Innovation and its Sustainability Potential

Product-Service Systems (PSS) are a particular type of value proposition that shifts the business focus from the design and sale of (physical) products alone, to the offer of a bundle of products and services that are jointly capable of satisfying a particular customer demand (UNEP 2002). In other words it is a value proposition oriented to satisfy users through the delivery of functions instead of products (e.g. thermal comfort instead of methane, having clean clothes instead of washing machines and powder, etc.). From a sustainability point of view the PSS concept is interesting because embodies the potential of decoupling economic value generation from material and energy consumption.

F. Ceschin, *Sustainable Product-Service Systems*, PoliMI SpringerBriefs,
DOI: 10.1007/978-3-319-03795-0_2,

Before describing in details the PSS concept and the associated benefits, drivers and barriers, let's have a look to a preliminary example: the Pay-per-Use solution, a PSS offer developed by Ariston, an Italian appliances producer. Here, rather than selling a washing machine, Ariston offers to clients the possibility to have clean cloths without owning the product. The payment is based on number of washes and includes the delivery of a washing machine at home, electricity supply (not directly paid by the customer), maintenance, and end-of-life collection. Basically, rather than the 'traditional' forms of sale, ownership, consumption and disposal, this value proposition is focused on delivering a particular satisfaction: 'having clean cloths'. And this satisfaction is delivered through a mix of products (owned by Ariston) and services. Why is this PSS concept promising in terms of sustainability? Because within this business model Ariston is economically incentivised in reducing as much as possible the washing machine energy consumption (in order to reduce operational costs and maximise profits), and in designing and providing long lasting, reusable and recyclable washing machines (in order to postpone the disposal costs and reducing the costs for the manufacturing of new washing machines).

This example can help us to better understand the key differences between *traditional sale* and *functional sale*. In the *traditional business logic* the profit centre is directly linked to material goods: the objective is to sale more products, auxiliary products and spare parts. In this case, as underlined by Vezzoli (2007), the producer has an economic interest in reducing material and energy consumption during the production phase, but at the same time it has no direct economic interest in limiting consumption during use, nor in reducing disposal impact or valorising the resulting waste; sometimes the producer is even interested in selling products with a short life span, with the aim of accelerating replacement. In the *functional economy*[1] the profit centre is tied to the number of 'functional units' delivered to customers (Stahel 1994). This means that the provider is paid per unit of function delivered and not per unit of product sold. In other words functional economy is oriented to satisfy consumers through the delivery of functions instead of products (e.g. mobility instead of cars, having clean clothes instead of washing machines and powder), and this can potentially create, as showed in the previous example, economic incentives for producers/providers to reduce and optimise the amount of material and energy consumption.

Using Stahel words (2001), the traditional economy might be called a *"river economy"*, because its success is measured as the monetary flow at the point-of-sale, which is directly linked to the flows of goods. The river economy has therefore a linear structure. Continuing with Stahel's thought, functional economy might be called a *"lake economy"*, because it measures its success in terms of asset management, which can be potentially linked to the revalorisation of existing

---

[1] The concept of functional economy was elaborated by Stahel. See: Stahel (1994, 1997a, b, 1998, 2001); Stahel et al. (2000), and Giarini and Stahel (1989/1993).

stocks of products and the optimisation of their utilisation. The lake economy is therefore structured in loops.

As underlined by Mont (2004a), functional sales change the entire meaning of value. In the traditional sale producers are seen as creators of value and customers as value destroyers, while in the functional economy producers become providers of value while customers become users of value. It becomes in the interest of both parties to make sure that the function is constantly fulfilled and the value is continuously provided. Moreover, in traditional sales there is a conflict between the interests of producer and customer, while in functional sales these interests can potentially be more aligned (White et al. 1999). In fact in traditional sales producers seek to sell more products, auxiliary products and spare parts, while customers try to reduce costs associated with these products (Mont 2004a). For the same reason producers are not interested that customers take good care of products, because neglect can lead to premature product obsolescence and, as consequence, to a new purchase. On the other hand functional sales can potentially align producers' and consumers' interests, creating incentives for producers/providers to reduce, at least partially, the aforementioned drawbacks (ibid.). In fact, if what is offered is a performance or the fulfilment of a satisfaction (e.g. having clean cloth), producers and providers will be interested in reducing the amount of material and energy inputs needed to deliver that performance. Thus, they could be potentially incentivised in creating the conditions to reduce the costs associated with the use, to extend products life spans and to facilitate refurbishing, components reuse and material recycling.

In conclusion functional economy (which is at the core of the PSS concept) represents a promising approach to re-orient the current structure of production and consumption towards a more dematerialised one. However, it must be highlighted that PSS is not synonymous of environmental sustainability: a PSS-oriented business model does not guarantee environmental benefits. As stated by UNEP (2002), PSSs only have the potential to do so. PSSs in fact represent a promising concept to move in the direction of sustainability; but this potential can be exploited only if PSSs are properly designed, developed and delivered.

Starting from these preliminary considerations the following text illustrates in details the concept of PSS, and its related drivers, barriers and potential benefits.

2.2 PSS Definition and Classification

2.2.1 PSS Definition and Main Characteristics

Several definitions have been given in the last decade to the concept of PSS (Table 2.1 lists the most important ones). As underlined by Mont (2004a), various authors put different emphasis on different priorities and consequently have different expectations on PSSs. Some authors define PSSs focusing only on the

Table 2.1 Definitions of product-service system

Author/s	Year	Definition
Goedkoop, van Halen, te Riele and Rommens	1999	A PSS (or combination of products and services) is a set of marketable products and services jointly capable of fulfilling a need for a client. [...] The PSS may lead to a benefit for the environment in connection with the creation of a (new) business
Mont	2001	PSS is a system of products, services, networks of actors and supporting infrastructure that continuously seeks to be competitive, satisfy customer needs and have a lower impact than traditional business models
Centre for sustainable design	2001	A PSS is a pre-designed system of products, supporting infrastructure and necessary networks that fulfil users' needs on the market, have a smaller environmental impact than separate product and services with the same function fulfilment and are self-learning
UNEP (Manzini and Vezzoli)	2002	A PSS is the result of an innovative strategy that shifts the centre of the business design and sale of products only (physical) to systems offering products and services that are jointly capable of satisfying a given application
Brandsotter et al.	2003	PSS is a product of material and intangible services designed and combined so that both jointly are able to satisfy a specific need of a user. In addition a PSS may reach sustainability targets
Wong	2004	A PSS may be defined as a solution offered for sale that involves both a product and a service element, to deliver the required functionality
Baines et al.	2007	PSS is an integrated offering of a product and a service that provides a value. Using a PSS offers the opportunity to decouple economic success from material consumption and thus reduce the environmental impact of economic activity
UNEP (Tischner et al.)	2009	A PSS is a system of products and services (and infrastructure), to jointly cope with the needs and demands of customers in a more efficient way with better value for both businesses and customers, compared to only offering products [...]. PSS can decouple the creation of value from the consumption of materials and energy and thus significantly reduce the environmental impact in the life cycle of traditional product systems

market and competitive aspects, without referring to the sustainability ones (UNEP 2002; Wong 2004). Some others include the sustainability dimension, defining PSSs as offer models with intrinsic lower environmental impact than traditional business models (Mont 2001; Centre for Sustainable Design 2001; Tischner et al. 2009), or as solutions which only have the potential to achieve sustainability improvements (Goedkoop et al. 1999; Brandstotter et al. 2003; Baines et al. 2007).

Basically, a PSS can be described as *an integrated system of products and services, delivered by one or more socio-economic actors, and designed to fulfil a specific customer need*. The word 'system' refers to both the *system* of products and services delivered to the client, and the *system* of actors that produces and offers the combination of products and services. Therefore, PSS is a specific type of value proposition whose emphasis is on satisfying customers through the delivery of functions rather than the selling of products or services. In order to add further clarity on the PSS concept, let us illustrate its key elements (Mont 2002a):

- *Products*: the tangible artefacts of the system.
- *Services*: they include services that make products available (sales services, renting, sharing, etc.), and services to manage products in the use and end-of-life phases (maintenance, upgrading, take back, etc.).
- *Network of actors*: it includes all the socio-economic actors needed to produce and deliver the PSS, and it comprises the partnerships and interactions between those actors belonging to that particular value chain or, as Normann and Ramirez define it, *"value constellation"* (Normann and Ramirez 1995).
- *Infrastructures*: they comprise existing collective and private systems (such as roads, communication lines, waste collection systems, etc.). PSS and infrastructures are strictly correlated: infrastructures affect the configuration of the PSS and at the same time the PSS can stimulate the development of new infrastructures or the modification of existing ones.

As previously underlined not all PSSs create environmental benefits. A PSS must be specifically designed, developed and delivered, in order to generate less material flows and emissions than the competing product orienting models. Which are the conditions needed to make this happen? First of all the PSS should be conceived in order to create economic and competitive incentives, for the actors of the value constellation, to lower and optimise material and energy consumption. Secondly, the products included in the PSS should be properly designed in order to exploit this potential: they should be designed with a low environmental impact in the various life cycle phases.[2] In other words the configuration of the PSS (in terms of stakeholders' interactions) constitutes the starting point towards achieving certain environmental results, but it is only with the proper design of the products associated with the PSS that these results can be actually achieved.

Starting from these considerations an *eco-efficient PSS* can be defined as a PSS *"where the economic and competitive interest of the providers continuously seeks environmentally beneficial new solutions"* (Vezzoli et al. 2013). It is a PSS in which the economic interest of the stakeholders involved in the PSS offer converges with an interest in optimizing the environmental resources consumption.

Of course, an eco-efficient PSS is not a sustainable PSS. The first one focuses on the economic and environmental dimensions, while the second one includes

[2] The design of these products should take in consideration *Life Cycle Design* (or *Eco Design*) criteria and guidelines. See Vezzoli and Manzini (2008), and Vezzoli et al. (2009).

also the socio-ethical dimension. Thus, a *sustainable PSS* can be defined as *a PSS where the economic and competitive interest of the providers continuously seeks environmentally beneficial new solutions, while maximising social well-being, equity and cohesion.*

2.2.2 PSS Classification

Most of the classifications proposed in the past distinguish between three main categories (Hockerts et al. 1993; Hockerts and Weaver 2002; UNEP 2002; Tukker and van Halen 2003; Tukker 2004; Vezzoli 2007; Vezzoli et al. 2013):

- *Product-oriented PSSs.* They can be defined as value offers where a company (or an alliance of companies/stakeholders) provides additional services to guarantee life cycle performance of the product (sold to the customer). These services can include for example maintenance, repair, upgrading, substitution, product take-back, etc. (UNEP 2002). This type of PSS reduces the user's responsibility in the use and/or disposal of the product/semi-finished product (owned by her/him), and can potentially drive the company's economic and competitive interest in continuously seeking environmentally beneficial new solutions, i.e. the economic interest could become something other than only selling a higher amount of products (Vezzoli 2007). In relation to this category it is possible to identify two more specific PSS types (Tukker 2004):
  - *Product-related services*, when producer/provider sells the product but also offers services that are needed during the use and or end-of-life phases. These services can include, for example, a financing scheme, a maintenance contract, an upgrading contract, a take-back agreement etc.
  - *Product-related advice/consultancy*, when producer/provider gives advices, in relation to the product sold, on the most efficient use (and/or disposal) of it. These services can include, for example, advices or courses on how to use the product.
- *Use-oriented PSSs.* They can be defined as value propositions where a company (or an alliance of companies/stakeholders) offers access to products, tools, opportunities or capabilities that enable customers to get the results they aim to. The client obtains the desired utility but does not own the product that provides that utility, and pays only for the time the product is actually used. Depending on the contract agreement, the user could have the right to hold the product/s for a given period of time (several continuous uses) or just for one use (UNEP 2002). The client thus does not own the products and does not operate on them to obtain the final satisfaction (the client pays the company to provide the agreed results). Again in this case the relation between the company and the client can potentially drive the company's economic and competitive interest to

continuously seek environmentally beneficial new solutions, e.g. to design highly efficient, long-lasting, reusable and recyclable products (Vezzoli 2007). Use-oriented PSSs can be divided into four main subcategories (Tukker 2004):

- *Product lease*, when producer/provider keeps the ownership of the product (and is often responsible for maintenance, repair and disposal). Customer pays a regular fee for an unlimited and individual access to the leased product.
- *Product renting or sharing*, when producer/provider keeps the ownership of the product and is responsible for maintenance, repair and disposal. Client pays for the use of the product (e.g. pay per hour) without having an unlimited and individual access. Other clients in fact can use the product in other moments (different users can sequentially use the product).
- *Product pooling*, when producer/provider keeps the ownership of the product and is responsible for maintenance, repair and disposal. Client pays for the use of the product (e.g. pay per hour) without having an unlimited and individual access. Differently from product lease and product renting and sharing, here the product can be used simultaneously by different users.
- *Pay-per-service unit*, when producer/provider keeps the ownership of the product and is responsible for maintenance, repair and disposal. Customer pays for the output of the product (e.g. pay-per-wash formula in relation to washing machines, pay-per-print formula in relation to printing machines, etc.). User can have a personal access to the product (e.g. in the Pay-per-Use solution, offered by Ariston, the washing machine is installed in the client's home), or a shared access (e.g. self-service washing centres). The difference with the functional result type is that in this case users have to operate on the product by themselves.

• *Result-oriented PSSs*. They can be defined as value offers where a company (or an alliance of companies/stakeholders) provides a customized mix of services (as a substitute for the purchase and use of products), in order to provide a specific 'final result' (in other words, an integrated solution to meet the customer's satisfaction). The mix of services does not require the client to assume (full) responsibility for the acquisition of the product involved. Thus, the producer maintains the ownership of the products and is paid by the client only for providing the agreed results. The customer benefits by being freed from the problems and costs involved in the acquisition, use, and maintenance of equipment and products (UNEP 2002). In other words, the client and producer/provider agree on a result, and there is no pre-determined product involved in the offer. The client does not own the products and does not operate on them to achieve the final satisfaction; the client pays the company to provide the agreed results. Again, the company can potentially have an economic and competitive interest in improving resource productivity, e.g. long-lasting, reusable and recyclable products (Vezzoli 2007). Within this category it is possible to identify two subcategories (Tukker 2004):

- *Activity management/outsourcing*, when a producer/provider outsources an activity to a third party, agreeing on performance indicators to control the quality of the outsourced activity (e.g. outsourcing of chemical management activity, or office cleaning).
- *Functional result*, when the producer/provider agrees with the client to deliver a final result (e.g. thermal comfort). There is no any predetermined product or technology involved: provider is free to decide the most effective means to deliver that result. Provider keeps the ownership of all products and equipment used to deliver the agreed result.

2.3 PSS Potential Benefits

There are several benefits that can result from an appropriately designed PSS. These can be grouped into three main categories: economic and competitive benefits, environmental ones, and socio-ethical ones.

Economic and Competitive Benefits The shifting towards a PSS-oriented business strategy can provide market opportunities and an improved strategic positioning (Goedkoop et al. 1999; Wise and Baumgartner 1999; Manzini et al. 2001; UNEP 2002; Mont 2002b). In other words PSSs can potentially improve the competitiveness of the company/ies (or more in general the alliance of stakeholders) involved in producing and delivering the offer. There are several reasons for this.

Firstly, a new PSS can represent a *differentiating offer* from the traditional product-based ones (Davidow and Uttal 1989; Kyj and Kyj 1994; Frambach et al. 1997), and can be more customisable to client wants and needs. Thus, moving up the value chain can represent an alternative to standardization and mass production (Baines et al. 2007), and a potential answer to the price competition from low-cost economies (Tukker and Tischner 2006a).

In addition a PSS can potentially provide an *added value to customers* compared to a product-based offer. Customers can in fact obtain the requested satisfaction without necessarily making large investments into products (Mont 2004a), and without being responsible for the costs and problems associated with the acquisition, use, maintenance and disposal of these products (UNEP 2002). PSS-oriented offers provide also increased flexibility to customers; the flexibility of the service element can in fact facilitate the customisation of the PSS in relation to the specific needs of single clients (Cook et al. 2006). This combination of advantages can stimulate the process of attracting new customers, which, as underlined by Mont (2004a) and Martinez et al. (2010), may be of strategic importance especially in mature industries and markets.

PSS can improve strategic positioning and competitiveness also because it can establish *longer relationships with customers* (Manzini et al. 2001; UNEP 2002; Mont 2004a). In fact the link between customers and the producer/provider does not end after the purchase choice (as it happens in the traditional product-based offers), but it continues during the whole period of the contract. Longer relations can potentially lead to *stronger provider/customer relationships* (UNEP 2002), increasing client fidelity. This longer and stronger connection can increase intensity of communication with customers, creating opportunities for feedback collection and consequently for the development of better value propositions, that in turn may increase customer loyalty (Mont 2004a).

The previously mentioned PSS high flexibility is also strategically important because it allows companies to be able to *respond more rapidly and easily to the changing market and customers preferences* (UNEP 2002). Moreover competitiveness is enhanced *because the combination of product and service elements is not easy to copy*, compared to single products (Dickson 1992; Ghemawat 1986; UNEP 2002).

Finally, PSS-oriented solutions can improve strategic positioning because they can help to *comply with existing and future environmental legislation* (such as Extended Producer Responsibility, environmental performance labelling, standards and specific international agreements, etc.), and to potentially do it in a way that adds value to customers and increases their satisfaction (Mont 2004a).

Environmental Benefits The environmental benefits of PSS-oriented solutions are related to the fact that there is a potential economic and competitive incentive, for the stakeholders involved in the PSS offer, to optimise the material and energy consumption. It is the same economic interest that pushes the actors involved in the PSS offer to improve environmental performances (White et al. 1999; Stahel et al. 2000; Heiskanen and Jalas 2000; Wong 2001; Zaring et al. 2001; UNEP 2002). This convergence between environmental and economic interests is defined eco-efficiency (Vezzoli 2007). Of course this eco-efficiency potential changes in relation to the specific characteristics of each PSS. However, generally speaking, the environmental benefits of an eco-efficient PSS are as follows (UNEP 2002):

- During the *use phase*, PSS providers have a potential economic interest to reduce the amount of resources consumed. In fact, if what is sold is a performance, fewer resources are used to deliver that performance and fewer costs will be sustained by providers, with a consequent profit increasing. Moreover, if PSS providers keep the ownership (or at least retains some responsibility) on the products over their life cycles, there is a further economic incentive to extend their life spans (in order to reduce maintenance and repair costs, and postpone the disposal costs and the ones for the manufacturing of new products).
- At the *end of a product's life*, PSS providers have the potential economic interest to re-use or re-manufacture the product/s and their components, in order to save on disposal costs and on the costs to manufacture new products.[3] For the same reason, providers are economically motivated to look into other ways to extend materials life, through recycling, energy recovery or composting.

[3] This happens when PSS providers keep the ownership, or at least retains some responsibility, on the product/s included in the PSS offer.

Furthermore recycling can also reduce the need to buy new raw materials, saving on manufacturing costs.

In summary, in eco-efficient PSSs producers/providers might be economically interested to:

- *extend the product (and its components') life span* (Vezzoli 2007). This includes facilitating maintenance, repair, upgrading and refurbishment of products, and reuse of components. In this way producers need to manufacture fewer products and components, and this leads to reducing raw material and energy consumption (Mont 2004a);
- *intensify product (and its components') use* (Vezzoli 2007), meaning that a (greater) number of people use the same product (or component) at different times (e.g. sharing and pooling schemes). A product used more intensely than others leads to a reduction in the quantity of products needed at a given time and in a given place to meet a customer demand (Vezzoli and Manzini 2008);
- *extend materials' life,* in order to valorise materials from scrapped products. Rather than ending up in landfills, these materials can be re-processed to obtain new secondary raw materials or incinerated (burned) to recover their energy content (Vezzoli 2007);
- *minimise the use of resources*, in order to reduce the usage of materials and energy of a given product in all its life cycle phases (Vezzoli 2007);
- *use of more advanced technologies,* both in products and in manufacturing stages. In fact the absence of ownership transfer to the customer at the point of sale makes providers not tied by the product price (as in the case of selling products) (Mont 2004a, b);
- *substitute obsolete products with new and more efficient ones* (Vezzoli 2007), if the cost of these new products can be more than offset by reduced operative costs (e.g. less costs in the use phase due to energy savings);
- *implement sufficiency solutions*, in order to organise operations so that the need for a product or service is reduced or eliminated without compromising consumer satisfaction (Stahel et al. 2000; Heiskanen and Jalas 2000);
- *educate customers* on how they could use the products so that they extract maximum utility with the least environmental impact possible (Mont 2004a).

Of course, as previously underlined, these potential reductions must be verified case-by-case, and balanced against the possible increase in costs of servicing, transportation, disposal and recycling. For example many PSSs are characterised by high transport intensity, and other environmental benefits in the system may not compensate this (Ellger and Scheiner 1997; Graedel 1998). Thus, a PSS must be specifically designed, developed and delivered, in order to generate less material flows and emissions than the competing product orienting models. Moreover the proper configuration of the PSS (in terms of stakeholders' interactions) constitutes only the starting point towards achieving certain environmental results. This may create the economic and competitive incentives for the actors to lower and optimise material and energy consumption, but in order to exploit this potential the products included in the PSS have to be properly designed (i.e. with a low environmental impact in the various life cycle phases).

Furthermore, even when properly designed, it has been observed that some PSSs could generate unwanted side effects, usually known as rebound effects (UNEP 2002). Society as a whole is a set of complex, inter-related systems. As a result, something may happen that turns potential environmentally-friendly solutions into increases in consumption of environmental resources at the practical level (ibid.). One example is the impact of PSS on consumer behaviour: if the client does not own the product he/she uses, he could be led to careless behaviours. Nevertheless, PSS development seen as a whole presents great potential for generating win-win solutions that promote profit and environmental benefits. It has the potential to provide the conditions to enable communities to leapfrog to less resource-intensive (more dematerialized) systems of social and economic systems (Vezzoli et al. 2013).

Socio-Ethical Benefits A PSS-oriented solution can provide several benefits concerning the socio-ethical dimension. They can be grouped in two main clusters: benefits for customers and for the society at a whole. Socio-ethical benefits for *customers* include the following:

- PSSs may focus much more on the needs and value of customers (thanks to their customisation potentials) and thus can *improve quality of life* (Tukker et al. 2006a). Moreover PSSs may involve customers directly in the development and customisation of the offer, increasing their satisfaction (ibid.).
- Use- and result-oriented PSSs does not require payment for the full value of the product, thus it is more accessible for consumers who could not afford to buy this product (ibid.). This means that they may represent a *significant opportunity for contexts with fewer economic possibilities* (i.e. low-income and emerging contexts) to respond more easily to unsatisfied social demands with lower overall costs (UNEP 2002).

For the *society*, socio-ethical benefits comprise the following:

- PSSs are more labour and relationship intensive, thus they can lead to an *increase in local employment and dissemination of competences* (UNEP 2002).
- PSS development, when it is based on existing products, can sometimes be done with limited investments. These PSSs may create *new business opportunities for entrepreneurs* (Tukker et al. 2006b). Again it must be stressed out that this may represent an important opportunity for contexts with fewer economic possibilities: i.e. low-income and emerging contexts.
- PSSs can *strengthen the role of local economy* because they are more focused on the context of use: services are created at the same time and often at the same place when and where they are consumed (Tukker et al. 2006a). For this reason PSSs could trigger a greater involvement of more local, rather than global, stakeholders, thus fostering and facilitating the reinforcement and

prosperity of the local economy (UNEP 2002). Thus they may empower and enhance local resources, by safeguarding, *regenerating and empowering local economies* and the related human and natural resources (Vezzoli 2007).

It has to be underlined that, similarly to the environmental dimension, PSS could also generate socio-ethical rebound effects. For example, some PSSs may deliver services that customers used to perform themselves; the result is that users could lose skills and become dependent to market parties (Tukker et al. 2006a). Another example is the activity management PSS type (ibid.): this can result in an outsourcing of activities with the main aim of using a low cost working force (e.g. relocating a telephone help desk service to a call centre in an emerging or low-income country). Thus, as in the case of environmental sustainability, PSSs have to be carefully designed in order to avoid socio-ethical unwanted effects.

2.4 PSS Drivers

In the previous sections we presented the concept of PSS, described its main typologies, and illustrated its main potential benefits. This section illustrates the main drivers to push companies (and other stakeholders) to shift towards PSS-oriented business strategies. Drivers are described according to two categories: drivers for companies and drivers for customers.

Drivers for Companies The main driver for companies is associated to the *need to find new market opportunities* and in general *increase the competitive position*. In mature economies companies are increasingly challenged by countries with a low-cost labour base (Tukker and Tischner 2006a; Baines et al. 2007), which is leading to sheer price competition and eventually to low profit margins. Thus, companies in many sectors recognised that it is not sufficient to compete on product quality and efficiency of operational and production processes (Mont 2004a). Moreover product-based companies have proved to be relatively easy to imitate by competitors, while PSSs are less easy to replicate (Dickson 1992; Ghemawat 1986; UNEP 2002). This has pushed many companies to recognise the need to move up the value chain and focus on the strategic integration of products and services as a source of sustainable competitive advantage and corporate profitability (Oliva and Kallenberg 2003; Cohen et al. 2006; Rosen et al. 2003).

In addition, the process of adoption of a PSS-based strategy is frequently incentivised by the companies' *need to differentiate their offers* to better respond to new customer demands (Mathieu 2001; Oliva and Kallenberg 2003; Gebauer et al. 2006). In short PSSs can potentially respond more appropriately to the current need to move from standardisation and mass production towards a mass-customisation strategy (UNEP 2002; Mont 2004a).

Moreover Mont (2002b, 2004a) states that important internal enabling factors in defining whether the company will explore possibilities of providing PSS-oriented solutions, and to what extent, are the *commitment of top management*, and the *presence of a 'catalyst'* (a person who internally can market and promote the concept). Mont, in her PhD research (2004a), also found out that for some companies the environmental issue, when linked to economic savings, represented an internal driver to initiate PSS-oriented businesses.

In relation to external drivers, the most significant one to push companies to change their businesses is the development of more *stringent environmental regulations* (Mont and Lindhqvist 2003; Ceschin and Vezzoli 2010). Pollution charges, fiscal incentives for pollution abatement, Extended Product Responsibility programmes, etc. can stimulate companies in improving their solutions in terms of environmental impact. The development of PSS-oriented businesses can represent an answer to this regulative pressure (Mont 2002b). Also, it can represent a strategy to deal with future regulations (James and Hopkinson 2002), and turn them into a competitive advantage (Agri et al. 1999). However, it must be stressed out that these policy measures encourage environmentally better products and services but do not necessarily steer companies towards the development of PSS-oriented innovations (Mont and Lindhqvist 2003; Ceschin and Vezzoli 2010).

Another important external driver is the *growing public concern about environmental and socio-ethical issues* (Mont 2002b, 2004a). There is in fact an increasing number of actors (ranging from citizens to public institutions, and including society at large) aware of these problems (Larsson et al. 1996). This should push companies to improve the environmental performance of their processes, products and services. However, it has to be underlined that the demand of more sustainable products and services is very often answered by companies with 'greenwashing' strategies or, using Lanzavecchia words, with *"the development of ecological simulacrums"* (Lanzavecchia 2000).

Drivers for Customers Mont (2002b) reports that *customers expect reduced risks and liabilities associated with handling the product.* PSS-oriented solutions can respond appropriately to this need, relieving customers from the responsibility for a product. However, this finds a confirmation especially in the Business to Business sector, while in the Business to Customer sector several barriers to ownerless consumption can be identified (see Sect. 2.5).

For private customers the main drivers to choose a PSS-based solution are inherent to the nature of the product: when products are expensive and not used very often, when maintenance costs are high and when products take up storage space, private customers tend to pay for services (Mont 2004a). In addition it has to be underlined that another driver is that ownerless-based solutions may also represent a certain status: let us think for example to cultural events or the use of a taxi.

2.5 PSS Barriers

Despite the aforementioned potential benefits and drivers it has to be underlined that the application of eco-efficient and sustainable PSSs is still very limited. An important reason is that this kind of business concepts requires, in most of the cases, substantial changes in existing customers habits, companies organizational structures and regulative frameworks. The following text illustrates in details these barriers.

2.5.1 Barriers for Companies (and in General for PSS Promoters)

Internal Mindset and Capabilities For companies the first barrier is related to the fact that the adoption of a PSS strategy is more complex to be managed than the traditional way of delivering products alone. For this reason there is the *need to embed a PSS culture within the organisation* (Martinez et al. 2010). In other words *changes in corporate mindset and organisation* are required in order to support a more systemic innovation and PSS-oriented businesses (UNEP 2002). Since the capabilities and knowledge for producing and selling products are considerably different than those of managing PSSs, it is clear that *companies require new competences, skills and experience,* in relation to both management and design activities. There is in fact the need of:

- structuring the organization in a way to be competent at designing, making, and delivering PSS offers (Baines et al. 2007);
- acquiring PSS design methods and tools, which companies can use to orientate, design and assess sustainable PSS concepts (UNEP 2002);
- acquiring life-cycle costing methods (ibid.);
- developing performance metrics to measure organisations' ability in effective and efficient delivery of PSS offers (Martinez et al. 2010);

For these reasons personnel need to be trained and perhaps even additional personnel should be recruited (Mont 2004a). It has to be underlined that these changes may also be hindered by *internal conflicts between business functions* (Stoughton et al. 1998; White et al. 1999; Fishbein et al. 2000). These conflicts may also be enhanced by the *absence of an internal common language and alignment of mindsets* (Martinez et al. 2010).

Another internal barrier is determined by the *changing of systems and sources of gaining profit* (Mont 2002a): PSS business models require in fact medium-long term investments compared to the short-term profit generated at the point-of-sale. Consequently PSSs are connected with uncertainties about cash flows (Mont 2002b), which leads producers to perceive PSS business more risky than product-based one (European Commission 2001). These changes in the sources of gaining profit require *new internal accounting systems* and the restructuration of financial functions (Mont 2004a).

A further obstacle is the difficulty of quantifying the savings arising from PSS in economic and environmental terms, in order to market the innovation to stakeholders both inside and outside the company, or to the company's strategic partners (UNEP 2002).

Relationship with Stakeholders Along the Value Chain PSS innovations require companies to *adopt a different approach with stakeholders along the value chain*. The development and delivering of PSSs need in fact the building up of a strong collaboration among these actors. This may be a barrier, because of the *fear of sharing sensitive information* about companies' processes, products and technologies (Mont 2004a). Another obstacle is that partnerships and entrepreneurial interdependence may lead to *reduced control of core competencies* and *reduced influence on business decisions* (UNEP 2002). In relation to the value chain another barrier is represented by the potential *conflict of interests* between companies that aim to reduce sales volumes of material products and traditional interests of retailers that aim to increase sales (Cooper and Evans 2000).

Relationship with Customers In relation to customers, the main barrier for companies is to *convince them that a PSS solution can better fulfil their needs and wants* (Mont 2004a). Middlemen need to be appropriately trained in order to accomplish this task (ibid.). Moreover the *definition of contracts and their negotiations* may be problematic, hindering the process of acquiring new customers (Martinez et al. 2010). A further barrier connected to customers is related to the fact that providers are often worried about the *reduced care for products* that customers could have if they do not own the products (Mont 2004a).

2.5.2 Barriers for Customers

Consumption behaviour is a matter of individual choice, but it is also influenced by social norms and institutional settings. The current and dominant consumption behaviours put several constraints to the diffusion of alternative sustainable PSSs. Let us summarize the most important factors that determine this opposition; we will follow Mont's line of thought (2004b), classifying them in two main categories: economic and socio-psychological factors.

Economic Factors From an economic perspective, Røpke (1999) states that current consumption behaviours are firstly determined by the entire history of industrial development. The Industrial Revolution led to increased production volumes and reduced product prices, determining the need to sell more and more new products. This in turn encouraged the creation of demand for all the produced artefacts, and therefore strategies were defined to boost consumption. In relation to this Kilbourne et al. (2001) state that economic and political institutions have persuaded people to believe that higher material prosperity is the expected behaviour.

Another cause that contributes to reinforcing material consumption levels is related to so-called externalities. Since environmental and social costs connected to products are not included in their market prices, it can become difficult for sustainable PSS solutions to compete with industrially produced products (Mont and Lindhqvist 2003; Ceschin and Vezzoli 2010). Moreover the cost of labour is increasing and therefore it can be cheaper for customers to buy product-based offers (e.g. washing machines) instead of labour-intensive solutions like PSS-based offers (e.g. clothing care services).

In addition it must be underlined that customers show a lack of knowledge and understanding about the PSS concept (Mont 2004a; Catulli 2012). This generates uncertainties related to unclear risks, costs and responsibilities, and can lead customers to difficultly understand the benefits of a PSS offer. Moreover, it is sometimes problematic for them to accept producers in the role of providers of services.

Many customers (in particular in relation to the Business to Consumer sector) also lack a general understanding about life cycle costs (White et al. 1999). For this reason it is sometimes not easy for them to understand the potential economic benefits of PSS-oriented solutions. PSS-based offers are in fact usually perceived by the end-user as more expensive if compared to the purchase of products (even if sometimes the contrary is true), since the total cost of ownership (including use, maintenance, repairs and disposal costs) is not taken into consideration in the purchase of a product.

Socio-Psychological Factors Economic studies are traditionally based on the assumption that consumers are rational decision-makers whose choices are driven by utility maximization, with price and income factors as most important in making choices. However, as underlined by Mont and Plepys (2008), consumer behaviour has been found to be far more complicated than merely a rational response to prices, being influenced by different internal and external drivers induced by human psychology, social norms and institutional settings.

Sociological studies underline the role of habits in influencing consumption behaviour, arguing that consumption choices are dependent on prior consumption patterns. In relation to sustainable PSSs, the problem is that solutions based on sharing and access contradict the dominant and well-established norm of ownership (Behrendt et al. 2003), making consumers hesitant to accept ownerless-based solutions (Goedkoop et al. 1999; Manzini et al. 2001; UNEP 2002). This is especially true for particular types of needs (e.g. for washing our clothes we are not accustomed to the idea of a washing machine in our home that does not belong to us), while in other cases ownerless-based solutions have entered into our routines (e.g. the use of public transport services). It has however to be underlined that, compared to private customers, business customers tend to prefer functional sales to product ownership (Alexander 1997; Stahel 1997a). Moreover, as underlined by Wong (2004), the diffusion of a PSS in the consumer market is highly dependent on being sensitive to the culture in which it will operate; in fact he observes that PSSs have been more readily accepted in communal societies like Scandinavia, the Netherlands, and Switzerland.

Another barrier to the diffusion of ownerless-based solutions is the fact that the quantity and quality of accumulated goods is perceived as a measure of success in life, because it represents an indicator of a certain position in society (Mont 2004b; Catulli 2012). Moreover, as underlined by Halkier (1998), the current trend towards individualization is boosting consumption demand, because a person's identity is no longer defined by a community but rather by the goods s/he owns (goods that represent the signals of one's own identity). In relation to this it must be stressed out that refurbished products and sharing schemes may be perceived as second-class status (Mont 2004a).

Mont (2004a) also points out that for certain PSS categories providers need to develop systems for monitoring and managing the products condition at customer sites. This entails entering customers' facilities and getting access to information about some of customers' activities. The privacy issue, for some customers, may be a sensitive one.

In addition, hesitation towards offers based on ownerless access and sharing can also be linked to the perception of independence, hygiene and intimacy usually connected to one's own products.

Some Windows of Opportunity Even if there are barriers that may hinder the acceptance of ownerless-based offers, it must be emphasized that there are also some windows of opportunity that can be exploited to favour the acceptance of such solutions (Mont 2004b). Firstly, while traditional economics argues that users demand physical products to satisfy their needs, the works of some sociologists (e.g. Max-Neef 1991) tells us that needs can be fulfilled by material and non-material *"satisfiers"*. Moreover material consumption is not linked to happiness; in fact more materialistic people are not always happier than less materialistic ones (Belk 1985; Max-Neef 1995). In addition, some studies state that an increase in consumption levels represents the need to satisfy psychological and social aspirations rather than material ones (Jackson and Marks 1999). On the same line of thought, Hacker (1967) argues that the purchase of the same brand represents a substitute for a lost sense of community. Moreover, in relation to goods possession, even if it is true that this is perceived as a measure of a certain social status, it can also be proposed that ownerless solutions may represent a certain status; let us consider for example the use of a taxi, access to education or cultural events (Mont 2004b).

2.5.3 Context-Related Barriers

The so called externalities and the high cost of labour in industrialised countries are the main context-related factors, which, as illustrated before, contribute to reinforce material consumption and hinder PSS-based solutions. Moreover, other context-related barriers to be overcome may also include a lack of external infrastructure and technologies (e.g. for product collection, remanufacturing or recycling) (UNEP 2002), and the difficulties faced by governmental institutions to create regulatory drivers to support the promotion and diffusion of this kind of innovations (Mont and Lindhqvist 2003; Mont 2004a).

2.5.4 A Matter of Institutions

As we have seen sustainable PSS innovations usually encounter the opposition of the existing customer habits, companies organizational structures and regulative frameworks. Thus, their introduction and scaling up require breaking down the routine behaviours that are daily reproduced by individuals, groups, business communities, policy actors and society at large. This everyday routine shapes and is shaped by social institutions (Giddens 1984). Institutions are broad concepts that comprise (Scott 1995): regulative and legislative constraints (*regulatory institutions*); societal norms, moral and ethical rules, and established behavioural patterns and lifestyles (*normative institutions*); and the learning processes that help individual to understand and interpret reality and support decision making processes (*cognitive institutions*). These institutions constitute a *"structure of rules"* that provides meaning and stability to social behaviours and drives the everyday routine activities (Giddens 1984). A structure that is continuously embedded in the course of the action: structure creates routine actions, which reinforce structure, which creates actions, and so on in a circular process, defined by Giddens *"structuration"*. In other words, when actions become routines the underlying structure is more and more embedded in the culture, and these actions become taken for granted by individuals and societal groups. Using Ehrenfeld's (2008) words *"we become more and more machinelike and less and less conscious of the structure that drives actions"*.

If anything changes in this structure of rules, individual and societal groups will reproduce their behaviour over and over. Thus, the main issue to be faced in the process of introducing and diffusing sustainable PSSs is to understand which strategies are the most effective to change the current and dominant institutions. In other words we must understand what stimulus could be used to make emerge new basic belief, habits, and routines capable to promote PSS-oriented solutions.

2.6 Sustainable PSS Design

The ideation and development of eco-efficient and sustainable PSSs requires moving from *product thinking* to *system thinking* (Manzini et al. 2001; Manzini and Vezzoli 2003). This is essential in order to breakdown the business as usual attitude. This new design attitude can be articulated as follows:

1. Firstly, a *"satisfaction-system" approach* is required (Vezzoli 2010). This means that the starting point is the satisfaction of a particular customer demand. The focus is on designing the combination of products and services associated with the fulfilment of a particular customer demand. In other words there is an enlargement of the design scope, from a single product or service to the set of products and services capable to fulfil a given demand of satisfaction.

2. Secondly, a *"stakeholder configuration" approach* is required (Vezzoli 2010). This means that the design approach should focus on designing not only the combination of products and services, but also the stakeholder network configuration required to produce and deliver the PSS offer. To better explain this approach it might be useful to draw a parallel with product design. In designing a product a traditional designer defines the technical and aesthetic characteristics of its components, and their connection systems. In a similar way a PSS designer must imagine innovative types of 'connections' (partnerships and interactions) between appropriate components (represented by the socio-economic stakeholders) of a system responding to a particular customer demand of satisfaction. In other words, designing the configuration of a system means understanding who are the best socioeconomic stakeholders (components) and what are the best interactions and relationships among them (connections).
3. Thirdly, a *"customer-oriented" approach* is required. The relationship between the customer and the actors producing and delivering the PSS plays a key role in the design of an effective PSS (Baines et al. 2007). In fact the early involvement of customer is essential to achieve a PSS offer that responds to customer wants and needs (Luiten et al. 2001). It is therefore suggested that customers should be treated as innovators, emphasizing a shift towards a value co-creation process, whereby professional customers and end-users play an organised and important role in the design process (Rocchi 2005; Luiten et al. 2001).
4. Finally, a *"system sustainability" approach* is required (Vezzoli 2010). As underlined in the previous sections, not all PSS innovations are eco-efficient and/or socially sustainable. Thus, it is crucial to appropriately design the stakeholder network configuration (offer model) in order to make stakeholders economically incentivised in improving the environmental and socio-ethical performance of the PSS.

Looking to the PSS design process from a more operational perspective, several methods and tools have been developed in the last years to support the designing of eco-efficient and sustainable PSSs.[4] The most relevant ones come from the results of recently funded European research projects.[5] These methods are usually organised around four main phases: *preparatory phase* (or strategic analysis), *exploring opportunities*, *PSS concept design* and *PSS engineering*.

---

[4] See for example: *Kathalys, method for sustainable product-service innovation* (Luiten et al. 2001); *DES, Design of eco-efficient services methodology* (Brezet et al. 2001); *MEPSS, Methodology for Product Service System development* (Van Halen et al. 2005); *MSDS, Method for System Design for Sustainability* (Vezzoli et al. 2009; Vezzoli 2010).

[5] *SusHouse* (Strategies towards the Sustainable Household 1998–2000), *ProSecCo* (Product-Service Co-design 2002–2004), *HiCs* (Highly Customerized Solutions 2001–2004), *MEPSS* (MEthodology for Product Service System development 2002–2005), and *SusProNet* (Sustainable Product Development Network 2002–2005).

A wide range of tools has been developed in association with these methods and can be used to support the whole PSS design process. These tools can be classified into the following broad categories: *tools to analyse and asses strengths and weaknesses*; *tools to support creativity and ideas generation*; *tools to steer the design process towards the development of sustainable PSSs* (tools to identify the sustainability design priorities, to support the ideas generation, to assess the sustainability improvements); *tools to support the visualisation of PSSs*; *tools to facilitate and stimulate co-design processes*. The methodological toolbox for eco-efficient and sustainable PSS development is fairly complete. The methods and tools developed so far can effectively support designers and companies in the process of conceiving eco-efficient and sustainable PSS concepts. However there are, especially in relation to the barriers illustrated in the previous section, some important gaps to be bridged:

- There is the need to better understand how PSSs can be designed in order to facilitate customer attraction, acceptance and satisfaction. In particular it is required to understand which factors influence customer satisfaction, and how it is possible to measure and evaluate this satisfaction (Baines et al. 2007). This knowledge would be extremely valuable at the design phase, in order to be integrated in the existing design methods.
- Available PSS design methods and tools put little (or no) emphasis on the implementation phase. Most of them do not even mention this phase, while other only provide general suggestions and guidelines. This is an important gap because, as previously said, the process of introduction and scaling up is often hindered by several barriers. Factors that influence the implementation process need to be understood and translated in design approaches, strategies and guidelines (Ceschin 2010, 2013a, b).
- Design knowledge, methods and tools need to be easily transferred to companies. Thus research results in this field need to be made available to companies and professional designers in a manner that encourages its use in practice. For example, in relation to this, Tukker and Tischner (2006b) suggest that the main challenge is to organise the available knowledge in an accessible way (including training and educational programmes), and develop an open case base including PSS concepts for different sectors.

Within this panorama of potential interventions, the next chapters aim to provide an answer to bridge the second gap.

References

Agri J, Andersson E, Ashkin A, Söderström J (1999) Selling functions: a study of environmental and economic effects of selling functions. CPM-report 1999:6. Chalmers Tekniska Högskola, Göteborg

Alexander M (1997) Managing the boundaries of the organisation. Long Range Plan 30:787–789

Baines T, Lightfoot H, Evans S, Neely A, Greenough R, Peppard J, Roy R, Shehab E, Braganza A, Tiwari A, Alcock J, Angus J, Bastl M, Cousens A, Irving P, Johnson M, Kingston J, Lockett H, Martinez V, Micheli P, Tranfield D, Walton I, Wilson H (2007) State-of-the-art in product service-systems. Proc Inst Mech Eng Part B J Eng Manuf 221:1543–1552

Behrendt S, Jasch C, Kortman J, Hrauda G, Pfitzner R, Velte D (2003) Eco-Service Development: Reinventing Supply and Demand in the European Union. Sheffield, UK: Greenleaf Publishing

Belk RW (1985) Materialism: Trait Aspects of Living in the Material World. J Consum Res 12: 265–280

Brandstotter M, Haberl M, Knoth R, Kopacek B, Kopacek P (2003) IT on demand—towards an environmental conscious service system for Vienna (AT). In: Proceedings of the 3rd international symposium on environmentally conscious design and inverse manufacturing: EcoDesign'03 (IEEE Cat. No. 03EX895), pp 799–802

Brezet H, Bijma AS, Ehrenfeld J, Silvester S (2001) The design of eco-efficient services. Methods, tools and review of the case study based "Designing eco-efficient Services" project. Report for Dutch Ministries of Environment (VROM). VROM, The Hague

Catulli M (2012) What uncertainty? Further insight into why consumers might be distrustful of product service systems. J Manuf Technol Manage 23:780–793

Centre for Sustainable Design (2001) Sustainable Service Systems (3S). Transition towards sustainability. In: Proceedings of the 6th international conference towards sustainable product design, De Rode Hoed, Amsterdam, 29–30 Oct 2001

Ceschin F (2010) How to facilitate the implementation and diffusion of sustainable product-service systems? Looking for synergies between strategic design and innovation sciences. In: Ceschin F, Vezzoli C, Zhang J (eds) Sustainability in design: now! Challenges and opportunities for design research, education and practice in the XXI century. Proceedings of the Learning Network on Sustainability (LeNS) conference, vol 1. Bangalore, 29 Sep–1 Oct 2010. Greenleaf Publishing, Sheffield, pp 440–454

Ceschin F (2013a) Critical Factors for managing the implementation and diffusion of eco-efficient product-service systems: insights from innovation sciences and companies' experiences. J Cleaner Prod 45:74–88

Ceschin F (2013b) The societal embedding of sustainable product-service systems. Looking for synergies between strategic design and transition studies. In: Vezzoli C, Kohtala C, Srinivasan A, Xin L, Fusakul M, Sateesh D, Diehl JC (eds) Product-Service System design for sustainability. Greenleaf Publishing, Sheffield (in press)

Ceschin F, Vezzoli C (2010) The role of public policy in stimulating radical environmental impact reduction in the automotive sector: the need to focus on product-service system innovation. Int J Automot Technol Manage 10:321–341

Cohen MA, Agrawal N, Agrawal V (2006) Winning in the aftermarket. Harvard Bus Rev 84:129–138

Cook M, Bhamra T, Lemon M (2006) The transfer and application of product service-systems: from academia to UK manufacturing firms. J Cleaner Prod 14:1455–1465

Cooper T, Evans S (2000) Products to services: a report for friends of the earth produced by the centre for sustainable consumption. Hallam University, Sheffield

Davidow WH, Uttal B (1989) Total customer service: the ultimate weapon. Harper & Row, New York

Dickson PR (1992) Toward a general theory of competitive rationality. J Mark 56:69–83

Ehrenfeld JH (2008) Sustainability by design: a subversive strategy for transforming our consumer culture. Yale University Press, New Haven, London

Ellger C, Scheiner J (1997) After industrial society: service society as clean society? Environmental consequences of increasing service interaction. Serv Ind J 17:64–579

European Commission (EU) (2001) Eco-services for sustainable development in the european community. European Commission, Brussels

Fishbein B, McGarry LS, Dillon PS (2000) Leasing: a step toward producer responsibility. Inform, USA

Frambach RT, Wels-Lips I, Gundlach A (1997) Proactive product service strategies. Industrial Marketing Management, 26:341–352
Gebauer H, Friedli T, Fleisch E (2006) Success factors for achieving high service revenues in manufacturing companies. Benchmarking, 13 (3):374–386
Ghemawat P (1986) Sustainable advantage. Harvard Bus Rev 64:53–58
Giarini O, Stahel W (1989/1993) The limits to certainty, facing risks in the new service economy (2nd ed). Dordrecht
Giddens A (1984) The constitution of society: outline of a theory of structure. Polity Press, Cambridge
Goedkoop M, van Halen C, te Riele H, Rommes P (1999) Product services systems, ecological and economic basics, Report 1999/36. VROM, The Hague
Graedel TE (1998) Lifecycle assessment in the service industry. J Ind Ecol 1:57–70
Hacker A (1967) A defence (or at Least an Explanation) of American materialism. Sales Management, March 1967, pp 31–33
Halkier B (1998) Miljbhensyn i Forbrug. Erfaringer og Forhandlinger i Ambivalente Hverdagsliv. Institut for Miljb, Teknologi og Samfund, Roskilde Universitets, Roskilde
Heiskanen E, Jalas M (2000) Dematerialisation through services: a review and evaluation of the debate. Ministry of the Environment, Environmental Protection Department, Helsinki
Hockerts K, Geissler F, Petmecky A, Seuring S. (1993) Kreislaufwirtschaft statt Abfallwirtschaft—Optimierte Nutzung und Einsparung von Ressourcen durch Öko-Leasing und Servicekonzepte [Closed loop economy instead of waste management—optimised use and resource saving through eco-leasing and service concepts]. Bayreuther Initiative für Wirtschaftsökologie, Bayreuth
Hockerts K, Weaver N (2002) Towards a theory of sustainable product service systems: What are the Dependent and Independent Variables of S-PSS? In: Proceedings of the INSEAD-CMER research workshop "Sustainable Product Service Systems: Key Definitions and Concepts", 9 May 2002
Jackson T, Marks N (1999) Consumption, sustainable welfare and human needs: with reference to UK expenditure patterns between 1954 and 1994. Ecol Econ 28:421–441
James P, Hopkinson P (2002) Service innovation for sustainability. A new option for UK environmental policy? Bradford University, Bradford
Kilbourne WE, Beckmann SC, Lewis A (2001) A multinational examination of the role of the dominant social paradigm in environmental attitudes of university students. Environ Behav 33:209–229
Kyj LS, Kyj MJ (1994) Customer service: product differentiation in international markets. Int J Phys Distrib Logistics Manage 24:41–49
Lanzavecchia C (2000) Il Fare Ecologico. Il Prodotto Industriale e i suoi Requisiti Ambientali. Paravia Bruno Mondadori Editori, Torino
Larsson R, Olsson-Tjarnemo H, Plogner AC, Ostlund S (1996) Market pull or legislative push: a framework for strategic ecological reorientation. Scand J Manage 12:305–315
Luiten H, Knot M, van der Horst T (2001) Sustainable product service-systems: the Kathalys method. In: Proceedings of the 2nd international symposium on environmentally conscious design and inverse manufacturing, Tokyo, 12–15 Dec 2001, pp 190–197
Mathieu V (2001) Product services: from a service supporting the product to service supporting the client. J Bus Ind Mark 16:39–58
Manzini E, Vezzoli C (2003) A strategic design approach to develop sustainable product service systems: examples taken from the 'environmentally friendly innovation' Italian prize. J Cleaner Prod 11:851–857
Manzini E, Vezzoli C, Clark G (2001) Product service systems: using an existing concept as a new approach to sustainability. J Des Res 1
Martinez V, Bastl M, Kingston J, Evans S (2010) Challenges in transforming manufacturing organisations into product-service providers. J Manuf Technol Manage 21:449–469
Max-Neef M (1991) Human scale development: conception, application and further reflections. The Apex Press, New York

Max-Neef M (1995) Economic growth and quality of life: a threshold hypothesis. Ecol Econ 15:115–118
Mont O (2001) Introducing and developing a product-service system (PSS) Concept in Sweden, IIIEE Reports 2001:6. Lund University and NUTEK, Lund IIIEE
Mont O (2002a) Clarifying the concept of product-service system. J Cleaner Prod 10:237–245
Mont O (2002b) Drivers and barriers for shifting towards more service-oriented businesses: analysis of the PSS field and contribution from Sweden. J Sustain Prod Des 2:89–103
Mont O (2004a) Product-service systems: panacea or myth? PhD dissertation, Lund University
Mont O (2004b) Institutionalisation of sustainable consumption patterns based on shared use. Ecol Econ 50:135–153
Mont O, Lindhqvist T (2003) The role of public policy in advancement of product service systems. J Cleaner Prod 11:905–914
Mont O, Plepys A (2008) Sustainable consumption progress: should we be proud or alarmed. J Cleaner Prod 16:531–537
Normann R, Ramirez R (1995) Le strategie interattive d'impres. Dalla catena alla costellazione del valore, ETAS libri, Milan
Oliva R, Kallenberg R (2003) Managing the transition from products to services. Int J Serv Ind Manage 14:160–172
Rocchi S (2005) Enhancing sustainable innovation by design. PhD thesis, Erasmus University, Rotterdam
Røpke I (1999) The dynamics of willingness to consume. Ecol Econ 28:399–420
Rosen LD, Karwan KR, Scribner LL (2003) Service quality measurement and the disconfirmation model: taking care in interpretation. Total Qual Manage 14:3–14
Scott WR (1995) Institutions and organizations. Sage Publications, Thousand Oaks
Stahel WR (1994) The utilisation-focused service economy: resource efficiency and product-life extension. In: Allenby BR (ed) The greening of industrial ecosystems. National Academy Press, Washington, pp 178–190
Stahel WR (1997a) The functional economy: cultural and organizational change. In: Richards (ed) The industrial green game. National Academy Press, Washington
Stahel WR (1997b) The service economy: 'wealth without resource consumption'? Philos Trans Royal Soc Lond 355:1309–1319
Stahel WR (1998) From products to services: selling performance instead of goods. Available on line: http://www.greeneconomics.net/Stahel%20Essay1.doc. Accessed 19 Aug 2011
Stahel WR (2001) Sustainability and services. In: Charter M, Tischner U (eds) Sustainable solutions: designing products and services for the future. Greenleaf Publishing, Sheffield, pp 151–164
Stahel WR et al (2000) From manufacturing industry to a service economy, from selling products to selling the performance of products. Product-Life Institute, Geneva
Stoughton M, Shapiro KG, Feng L, Reiskin E (1998) The business case for EPR: a feasibility study for developing a decision-support tool. Tellus Institute, Boston
Tischner U, Ryan C, Vezzoli C (2009) Product-service systems. In: Crul MRM, Diehl JC (eds) Design for sustainability: astep-by-step approach. UNEP, Paris
Tukker A (2004) Eight types of product-service system: eight ways to sustainability? Experiences from Sus-ProNet. Bus Strateg Environ 13:246–260
Tukker A, Tischner U (2006a) Product-services as a research field: past, present and future. Reflections from a decade of research. J Cleaner Prod 14:1552–1556
Tukker A, Tischner U (eds) (2006b) New business for Old Europe: product services, sustainability and competitiveness. Greenleaf publishers, Sheffield
Tukker A, Tischner U, Verkuijl M (2006) Product-services and sustainability. In: Tukker A, Tischner U (eds) New business for Old Europe. Product services, sustainability and competitiveness. Greenleaf publishers, Sheffield
Tukker A, van den Berg C, Tischner U (2006) Product-services: a specific value proposition. In: Tukker A, Tischner U (eds) New business for Old Europe. Product Services, Sustainability and Competitiveness. Greenleaf publishers, Sheffield

Tukker A, van Halen C (eds) (2003) Innovation scan product service combinations, manual. TNO-STB, Delft, and PricewaterhouseCoopers, Utrecht

United Nations Environmental Programme (UNEP) (2002) Product-service systems and sustainability: opportunities for sustainable solutions. UNEP, Paris

United Nations Environmental Program (UNEP) (2009) Design for sustainability: astep-by-step approach. UNEP, Paris

Van Halen C, Vezzoli C, Wimmer R (eds) (2005) Methodology for product service system. How to develop clean, clever and competitive strategies in companies. Assen, Van Gorcum

Vezzoli C (2007) System design for sustainability: theory, methods and tools for a sustainable "satisfaction-system" design. Maggioli Editore, Rimini

Vezzoli C (2010) System design for sustainability: theory, methods and tools for a sustainable "satisfaction-system" design (2nd ed). Maggioli Editore, Rimini

Vezzoli C, Manzini E (2008) Design for environmental sustainability. Springer, London

Vezzoli C, Ceschin F, Cortesi S (2009) Metodi e strumenti per il life cycle design: come progettare prodotti a basso impatto ambientale. Maggioli Editore, Rimini

Vezzoli C, Kohtala C, Srinivasan A, Xin L, Fusakul M, Sateesh D, Diehl JC (eds) (2013) Product-service system design for sustainability. Greenleaf Publishing, Sheffield (in press)

White AL, Stoughton M, Feng L (1999) Servicizing: the quiet transition to extended product responsibility. Tellus Institute, Boston

Wise R, Baumgartner P (1999) Go downstream e the new profit imperative in manufacturing. Harvard business review 77

Wong M (2001) Industrial sustainability (IS) and product service systems (PSS): a strategic decision support tool for consumer goods firms. University of Cambridge, Cambridge

Wong M (2004) Implementation of innovative product service-systems in the consumer goods industry. PhD dissertation, Cambridge University

Zaring O, Bartolomeo M, Eder P, Hopkinson P, Groenewegen P, James P, de Jong P, Nijhuis L, Scholl G, Slob A, Örninge M (2001) Creating eco-efficient producer services. Gothenburg Research Institute, Gothenburg

Chapter 3
Introducing and Scaling Up Sustainable Product-Service Systems: Insights from Transition Studies

Abstract Building upon theories and concept from transition studies, this chapter illustrates how radical innovations take place, and which are the related dynamics, processes and influencing factors. The chapter then illustrates the concept of socio-technical experiment and in particular its role in triggering transition processes. The text then discusses to which extent the concepts and insights from transition studies are valuable for the specific characteristics of sustainable PSSs. As a result, the chapter puts forward a conceptual framework to describe how the implementation and diffusion of sustainable PSS innovations take place, and explains, in a coherent way, the main influencing factors (and how they are interrelated). The argumentation is accompanied by the illustration of case studies.

Keywords Product-service system · Sustainable development · System innovation · Transition studies · Multi-level perspective · Strategic niche management · Transition management · PSS implementation · PSS commercialisation · Scaling up

3.1 Clarifying the Concept of System Innovation

3.1.1 Defining System Innovations

System innovations can be viewed as major shifts in the way societal functions (such as transportation, communication, housing and feeding) are fulfilled (Geels et al. 2004); they therefore refer to the change process, or transition, from one socio-technical system to another. Thus, the concept of system innovation is strictly linked to the concepts of *transition* and *socio-technical system*. Let us spend a few words on these concepts.

Transition refers to "*a change, passage or movement from one state or stage to another*" (Collins English Dictionary), and to the period of time during which the

F. Ceschin, *Sustainable Product-Service Systems*, PoliMI SpringerBriefs,
DOI: 10.1007/978-3-319-03795-0_3, © The Author(s) 2014

change takes place. Transitions can of course occur on different levels (e.g. transitions at the level of economic systems, at the level of organisations and firms, etc.). However, here we are interested on a specific type of transition: transition at the level of societal functions. Thus, using the words of Rotmans et al. (2000), transition is meant here as "*a continuous process of societal change, where the character of society (or of one of its complex subsystem) undergoes structural changes*".

As previously said, transitions imply changes in socio-technical systems. **Socio-technical systems** can be defined as the elements (and the linkages between these elements) needed to fulfil a societal function (Geels 2004a). They include technology, regulations, markets, user practices and habits, cultural meanings, policies, infrastructures, maintenance networks and supply networks (ibid.). Thus, the perspective on socio-technical systems is wider than for example the one on industry structures, and includes a broad variety of societal groups: e.g. firms, industries, users, public authorities, research centres, NGOs, etc.

In summary, the transition to one socio-technical system to another (system innovation) implies a process of change of both the structure of the system and the relationship among the actors in the system (Quist 2007). Moreover, it takes place in the spheres of production, distribution and consumption, and affects the technical and societal/behavioural dimensions (Elzen et al. 2004). In particular, as pointed out by Rotmans and Loorbach (2010), system innovations involve a fundamental change in:

- **structure**: the *physical structures* (infrastructure, technologies, resources, materials), *institutional structures* (rules, regulations, power structures) and *economic structures* (market, financing, consumption, production). Changes in structure comprise changes in how actors organise the things they do, either physically, institutionally or economically;
- **culture**: the sum of shared images, norms and values that together constitute the perspective from which actors think and act. Changes in culture comprise shifts in thinking, mental models and perceptions;
- **practices**: the sum of activities (routines, behaviour). Changes in practices comprise changes in what actors actually do, how they work or behave.

The characterisation of system innovations as multi-actors processes combining both technological and non-technical changes agrees with the descriptions provided by Quist and Vergragt (2000), who state that system innovations are "*comprehensive innovations with a long time horizon, requiring* (1) *the efforts of many stakeholders, and* (2) *a change of perspective and a cultural shift among these stakeholders*". From what it has been said, it is possible to summarise that system innovations are characterised by being:

- **multi-actor**: they involve a wide range of actors, including firms, industries, users, public authorities, governments, research centres, NGOs, etc. (Quist and Vergragt 2000; Elzen et al. 2004);

- **multi-factor**: they are not caused by a change in a single factor but are the result of the interplay of many factors that influence each other; they are a combination of technical, regulatory, societal and behavioural change (Elzen et al. 2004);
- **multi-level**: they implies changes at various levels: the micro-level of *niches* (new developments that in the beginning do not fit with an existing system), the meso-level of *socio-technical regimes* (which refer to dominant sets of culture, practices and institutions related to a specific field), and the macro-level of the *socio-technical landscape* (which comprises the wider societal, economic and political context in which actors interact) (Geels 2002a, 2005a);
- **highly uncertain**: they are difficult to be predicted and managed because of their complexity and their inherently highly degree of uncertainty (van den Bosch et al. 2005);
- **long-term processes**: they take relatively very long time to occur because they require multidimensional changes (Elzen et al. 2004).

3.1.2 System Innovations and Sustainable PSS Innovations

In the introduction of this chapter and in the previous chapters it has been said that PSS innovations can be considered system innovations. After having clarified the concept of system innovation it must be stressed out that not all sustainable PSS innovations are system innovations. There are two main reasons for this.

The first reason is related to the quality of change. Some PSS innovations cannot be considered system innovations because they do not bring substantial changes (e.g. in the societal/behavioural and regulative dimensions). For example *product-oriented PSS* (see Sect. 2.2.2) only entails changes in the relationship between the company and the customer, while it does not imply changes on other dimensions. In general it is possible to state that sustainable *product-oriented PSSs* usually cannot be considered system innovations. On the other hand sustainable *use-oriented PSSs* and *result-oriented PSSs* usually require multi-dimensional changes (e.g. changes in existing customer and user practice, companies organizational structures, regulative frameworks, culture, etc.), and therefore can be considered system innovations.

The second reason is related to the dimension of change or, in other words, to the degree of diffusion of the PSS innovation. If a sustainable PSS innovation is implemented only by few companies the effects in terms of changes in the socio-technical system are limited. In this case we cannot talk about a system innovation, rather about a niche innovation that has the potential to become a system innovation. In other terms a PSS innovation cannot be considered a system innovation until it becomes part of the mainstream way in which a societal need is fulfilled.

3.2 System Innovation Dynamics: The Multi-level Perspective Model

After having clarified the concept of system innovation, the focus of this section is on illustrating how system innovations come about and which are their dynamics. An often-used model for simplifying the sheer complexity of such long-term and multi-actor processes is the **Multi-Level-Perspective (MLP)** model. In the MLP, the dynamics of transitions depend on the interactions between three different levels (Geels 2002a, 2004a, 2005a): the **socio-technical regime** (meso level), the **niche** (micro level), and the **landscape** (macro level).[1]

3.2.1 Socio-technical Regimes, Niches and Landscape

Socio-technical Regimes The socio-technical regime can be understood as the dominant way of innovating, producing, distributing and consuming. Geels (2004a) defines it as a dynamically stable set of rules carried out by different social groups. These *rules* provide orientation and coordination to the activities of these *social groups*, determining the relatively stability of *socio-technical systems*. Thus, social groups, rules and socio-technical systems can be considered the three analytical dimensions constituting a socio-technical regime (ibid.). In particular:

- ***Social groups*** refer to a broad variety of actors: firms, industries, users, public authorities, research centres, NGOs, etc. Socio-technical systems are actively created and maintained by all these social groups. Their activities reproduce the elements and linkages in socio-technical systems. Geels (2005a) underlines that these social groups have relative autonomy but, on the other hand, they are also interdependent and interacting with each other.
- ***Rules***, or institutions, refer to a shared structure (e.g. values, routines, norms, protocols etc.) that guide behaviour of actors and social groups. Scott (1995) distinguishes three dimensions of rules: *regulative rules*, which refer to explicit formal rules, which constrain the behaviour of actors and social groups and regulate their interactions; *normative rules*, which refer to value, norms, moral and ethical rules, role expectations, duties, rights, responsibilities; *cognitive rules*, which refer to the frame through which meaning or sense is made: they help individual and groups to understand and interpret reality and support decision making processes.
- ***Socio-technical systems*** refer to the elements (and the linkages between these elements) needed to fulfil a societal function, and they include technology, regulations, markets, user practices and habits, cultural meanings, policies,

---

[1] The conceptualisation of the MLP model by Geels builds upon previous work (Kemp 1994; Schot et al. 1994; Rip and Kemp 1998; Kemp et al. 1998; Van den Ende and Kemp 1999).

infrastructures, maintenance networks and supply networks (Geels 2004a). Socio-technical regimes are characterised by being relatively stable and resistant to change. First, because rules and institutions guide regime actors in a specific direction discouraging the development of alternatives (ibid.). Second, because actors and social groups are embedded in interdependent networks (ibid.), and therefore they are resistant to major changes because they develop webs of interdependent relationships with other actors and social groups (Tuschman and Romanelli 1985). Third, because socio-technical systems have a certain 'hardness' which makes them difficult to change (Geels 2004a); in fact, once technical systems have been created they are not easily abandoned (Walker 2000). This leads to path dependence in socio-technical systems (Geels 2004a). In other words this resistance to change results in particular paths or trajectories. These trajectories are not only related to technology but also to other dimensions, such as policy, science, culture, market, etc. These trajectories are usually aligned and go towards similar directions, and this creates stability and resilience. This stability is however, of a dynamic kind; this means that innovation still takes place but is of an incremental nature (Geels 2004b). However, the activities of the different social groups and the resulting trajectories may at times diverge, leading to maladjustments and tension (Freeman and Louça 2001), and instability in socio-technical regimes (Geels 2004b).

Niches can be described as protected spaces that are 'isolated' from the influence of the dominant regime, where radical innovations can be tested and nurtured, become more mature, and potentially challenge and change regime practices and institutions. Thus, if socio-technical regimes generate incremental innovations, radical innovations are (usually) developed in niches. Niches are protected spaces (protection can come for example from subsidies, tax exemptions, strategic investments by companies, etc.) that shield innovations from mainstream market selection. Because of this shielded environment, niches act as "incubation rooms" for radical novelties (Schot 1998; Kemp et al. 1998). Niches provide protection because the selection criteria are different from those in use in the socio-technical regime (Geels 2004b). In sum niches are important because they provide locations for (Schot 1998; Kemp et al. 1998; Hoogma et al. 2002):

- experimenting and developing radical innovations deviating from the rules in the existing socio-technical regime;
- learning processes, e.g. about technical specifications, user preferences, public policies, etc.;
- building the social networks to support the innovations.

As underlined by Geels (2004a) the three analytical dimensions (rules, actors and socio-technical systems) also apply to niches, but the difference is the degree of stability. In fact in niches not all rules have become stable. There may be a high uncertainty about the best design, user preferences, public policy, etc. The same can be said about the configuration of the social network, which tends to be in flux (no clear role and relationships, some actors may join the process while other may leave, etc.) (ibid.).

Landscape The landscape is the relatively stable social, economic and political context in which actors interact and regimes and niches evolve (Rip and Kemp 1998). It represents the background for regimes and niches. It includes a set of heterogeneous and slow changing factors such as cultural and social values, socio-economic developments, demographic trends, political and international developments, etc., but also shocks and surprise events such as wars, rapidly rising oil prices or environmental disasters (Geels and Kemp 2000). Landscape can influence the regime and the niches, but cannot be influenced by them (at least in the short term). In other words the landscape is an external context in niches and regimes: while actors can change (to some extent) regimes, it is more difficult for them to change the landscape (it is beyond the direct influence of actors and social groups) (Geels 2004b).

The three levels are characterised by a different kind of structuration of activities. This level of structuration increases from niches, to regimes and landscape (Schot et al. 1998; Geels 2002a): in niches, experimentation activities go in many directions and social networks are precarious; in regimes, rules are stable and there is coordination in the activities of actors; in the landscape there is an even stronger structuration, characterised by widely shared cultural belief and values from which it is hard to deviate. The relation between the three concepts can be understood as a nested hierarchy (see Fig. 3.1): regimes are embedded in landscapes, and niches within regimes (Geels 2002a). This nested character also means that radical changes are not the result of dynamics at any specific level, but occur as linkages and interactions between the three different levels (ibid.). For example dynamics at the landscape level can put pressure on regimes and result in changes at the regime level (e.g. the establishment of new policies or the emergence of new user practices); changes at the regime level may result in decreasing stability and uncertainty (Raven 2004), and may create opportunities for novelties (Geels 2002a); on the other hand interactions can also go in the other direction, e.g. when radical novelties are developed in specific application domains and invade mainstream markets (Raven 2005).

3.2.2 Dynamics in Socio-technical Transitions

If socio-technical regimes stay stable and aligned radical innovations have few chances to challenge regime practices and remain pinned in particular niches (Rip and Kemp 1998; Kemp et al. 2001; Geels 2002a). Radical innovations can break from the niche level if the external conditions are right: that is to say when on-going processes at the level of regimes and landscape produce 'windows of opportunity' (ibid.). Geels (2004a) argues that there may be different causes determining these misalignments and tensions:

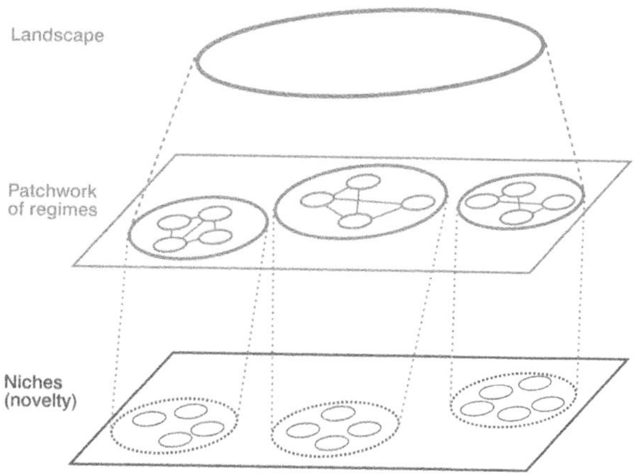

Fig. 3.1 Multiple levels as a nested hierarchy. Adapted from Geels (2002a)

- *Changes on the landscape level* may put pressure on the regime and cause internal restructuring (Burns and Flam 1987). An example is climate change, which is putting pressure and is stimulating changes on different sectors (e.g. energy, transport and agriculture ones).
- *Internal technical problems* in existing socio-technical regimes may contribute to stimulate actors to explore new technical directions.
- *Negative externalities and effects* (e.g. health risks, environmental impacts, etc.) may create pressure on the regime. As underlined by Van de Poel (2000) outsiders (such as societal pressure groups, outsiders firms, NGOs, etc.) play a crucial role to get negative externalities on the technical agendas of regime actors.
- *Changing in users preferences* (because of e.g. concern about externalities, cultural changes, policy measures such as taxes, discovery of new solutions, etc.) may lead to tensions in the regime and to the creation of new markets.
- *Strategic and competitive games between companies* may lead to open up the regime. For example firms may decide to support a particular niche if they think it can bring strategic advantages.

Using the MLP model, transitions can be conceptualised as follows (Geels 2002b) (see Fig. 3.2).[2]

Changes on the landscape level may put pressure on the regime. This pressure, combined with a regime internal destabilisation, could bring to a misalignment on the functioning of the regime and create windows of opportunities for radical novelties. At the same time, in niches, small networks of actors may test, incubate and support radical novelties on the basis of their expectations. In the niche, actors and societal groups learn about radical innovations, not only in terms of technical aspects but also about user preferences, regulations, policies, etc. Initially this

[2] Joore (2010) has combined this model with a prescriptive design process, developing the Multi Level Design (MLD) model, which clarify the mutual relationship between new products and societal change processes.

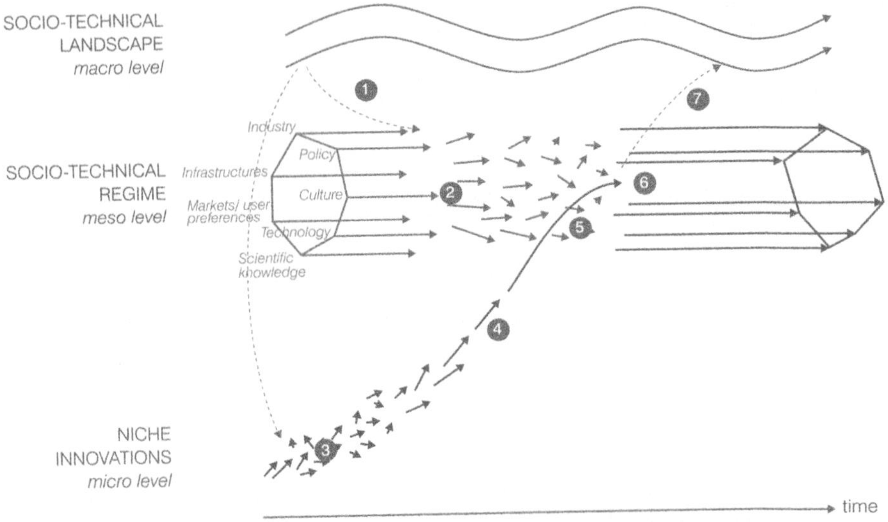

Fig. 3.2 The dynamic of transitions. *1* Changes on the landscape level may put pressure on the regime. *2* This, combined with a regime internal destabilisation, could bring to a misalignment on the functioning of the regime and create windows of opportunities for radical novelties. *3* Small networks of actors test, incubate and support novelties on the basis of expectations and visions. *4* Novelties get stable and internal momentum increases. *5* Novelties can take advantage of windows of opportunities. *6* Changes in the socio-technical regime take place. *7* New regime influences landscape. Adapted from Geels (2002b)

effort goes in many directions: there is much uncertainty about design, and this leads to a variety of different designs. Continuous experimentation and interactions between niche actors may lead to a broad community of actors who exchange experience, best practices and findings. This can lead to gradually stabilise radical innovations into a dominant design. The internal momentum increases and novelties can take advantage of windows of opportunities. Once the innovation breaks through into mass market it start to challenge the existing regime. This may lead to wide changes on different dimensions of the socio-technical regime. The new regime may influence wider landscape developments.

3.3 Socio-technical Experiments and Their Contribution in Triggering Transitions

3.3.1 Defining Socio-technical Experiments

The previous sections have showed the crucial importance of socio-technical experiments in potentially contributing to (learning about) a desired transition. These real-life context experiments represent strategic opportunities to develop and bring to mature radical (and therefore highly risky) innovations without the direct pressure of the mainstream market selection environment (Kemp et al. 1998; Hoogma et al. 2002; Brown et al. 2003; Van der Laak et al. 2007; Raven et al. 2010; Van den Bosch 2010). In this sense experiments can be used to:

- learn and improve the innovation on multiple dimensions, such as the technical, economic, market demand, user acceptance, political, regulative, environmental, and cultural ones;
- test and develop new institutional configurations (new routines, behaviours, interpretive frames, values, norms, etc.);
- build up a proper social network to support the innovation.

Several concepts referring to socio-technical experimentation have been elaborated in the last years. The most diffused ones are: *social experiments* (Verheul and Vergragt 1995), *experiments in Strategic Niche Management* (Kemp et al. 1998; Hoogma 2000), *transition experiments* (Rotmans et al. 2000; Van den Bosch 2010), *bounded socio-technical experiments* (Brown et al. 2003), and *experiments in Conceptual Niche Management* (Hegger et al. 2007).

Even if each concept presents its own peculiarities,[3] a socio-technical experiment can be described as a partially protected environment where a broad network of actors can learn and explore (1) how to incubate and improve radical innovations and (2) how to contribute to their societal embedding.

Its main characteristics are as follows (Ceschin 2012, 2013a):

- Firstly, **experiments are conducted with radical innovations**: innovations that require substantial changes on various dimensions (socio-cultural, technological, regulative and institutional).
- Secondly, experiments are not simple tests undertaken inside a company's laboratory but are implemented in **real life settings**. The idea is that only this kind of experience, outside the R&D settings, can truly lead to testing and improving radical innovations. Moreover these experiments take place at a small scale but strive to trigger changes at a wider scale.
- Thirdly, these experiments do not include only the actors more strictly linked to the innovation (such as producers, partners and suppliers). Instead, **a broad variety of actors is involved**, including also users, policy-makers, local administrations, NGOs, consumer groups, industrial associations, research centres, etc. In other words the aim is to recreate a whole socio-technical environment in a small scale. In this sense these experiments are characterised by a broad participatory approach (i.e. a variety of actors is involved in discussing, negotiating, co-creating and developing the innovation).
- Fourthly, the experiment is implemented in a **space protected from the mainstream selection environment**. The idea is to temporarily shield the innovation from the selection pressure (which consists of markets and

[3] For a comparison between the various concepts of socio-technical experiments see Ceschin (2012: pp. 88–94).

institutional factors), creating an alternative selection environment. There are different forms of protection: financial protection (such as strategic investments by companies, tax exemptions, and investment grants) and socio-institutional protection (such as the adoption of specific regulations). The crucial dilemma of protection measures is to find the right balance between the need to nurture the innovation and the need to prepare it for the selection pressures of a market environment (Weber et al. 1999).

- The aim of these experiments is to **learn about and improve the innovation on multiple dimensions**, not only the technical, economic, market demand and usability aspects, but also the political, regulative, environmental, cultural and social dimensions. In this sense the innovation is maintained open to continuous adjustments and refinements. In general experiments can also serve to identify the various resistances and barriers (institutional, regulative, economic, etc.) that can potentially hinder the future implementation and diffusion and understand how to address them.
- Moreover, and this is a crucial aspect, socio-technical experiments are not only aimed at testing and improving the innovation, but also at **stimulating changes in the socio-technical context**, in order to create the most favourable conditions for the innovation. In other words experiments are also strategically used to influence contextual conditions in order to favour and hasten the societal embedding process (for example, by influencing local administrations to adopt policy measures that support the innovation, or stimulating potential users to change their behaviours and routines).

3.3.2 Mechanisms Through Which Socio-technical Experiments Can Contribute to Transitions

The previous sections showed the crucial importance of socio-technical experiments to potentially contribute to transitions, and provided an overview on the main characteristics of socio-technical experiments. At this point the question is: through which mechanisms does a socio-technical experiment contribute to a transition? Van den Bosch, in her PhD thesis, identifies three main mechanisms through which socio-technical experiments can contribute to transitions (Van den Bosch 2010): **deepening**, **broadening**, and **scaling-up**.[4, 5] It must be stressed out that these mechanisms are not related in a sequential or chronological way, but can act upon an experiment simultaneously.

[4] The mechanisms deepening, broadening and scaling-up were first described in Rotmans and Loorbach (2006) and elaborated by Van den Bosch and Taanman (2006) and Van den Bosch and Rotmans (2008).

[5] Of course not all socio-technical experiments contribute to these three mechanisms. Experiments have to be properly conceived and managed in order to do so.

Deepening means learning as much as possible about an innovation within a specific context. Deepening enables actors to learn about local shifts in culture (ways of thinking, values, reference frameworks, etc.), practices (habits, ways of doing things, etc.) and institutions (norms, rules, etc.). In other words experiments are used to test and improve the innovation on multiple dimensions (not only the technical and economic ones, but also for example the cultural, regulative and institutional ones).

It has also to be underlined that within a socio-technical experiment learning is characterised as contextual (Van den Bosch and Taanman 2006): the same experiment in another context with different actors would possibly lead to different outcomes. As a consequence learning process within an experiment is partial because what is learnt is limited to the specific contextual conditions (Van den Bosch 2010).

The result of deepening is the development and reinforcement of the deviant set of culture, practices and institutions related to the innovation.

Broadening The second important mechanism is broadening, that means replicating the innovation in different contexts and linking it to other projects and initiatives. Since learning within an experiment is limited, experiments should be repeated in other contexts, in order to learn about different designs in different settings.

Broadening is related to the idea that different experiments, carried out simultaneously, can build on each other and gradually reinforce themselves (Raven 2005; Geels and Raven 2006). Within this perspective it is also important to strength synergies with other local similar projects and initiatives.

In this respect Meroni (2008) and (Jégou 2011) speak about "*synergizing*" or "*acupunctural planning*", a set of synergic self-standing local experiments that, adopting as a metaphor the practice of the traditional Chinese medicine, aim to generate changes in large and complex systems operating on some of their *sensible nodes*.[6]

Through processes of broadening the deviant set of culture, practices and institutions is (1) tested and extended to a variety of contexts and (2) linked to other existing projects and initiatives.

Scaling-up, the third important mechanism, means to embed the innovation in dominant ways of thinking, doing and organising. Scaling-up relates the socio-technical experiment to the regime. It is a process in which the innovation gains more influence and stability and gradually become part of the dominant way in which a societal need is satisfied. Scaling up takes place in many intermediate steps through which initially small changes in niches can eventually grow to broader changes in the dominant culture and practices of the regime (Van den Bosch and Rotmans 2008). In other words scaling-up relates to moving the innovation (and its initially deviant socio-technical practices) from a local experimental level to a mainstream level.

---

[6] Example of such kind of interconnected and synergetic experiments can be found in the *Feeding Milano* project, aimed at prototyping and implementing a network of "0 miles" food related services between the city of Milan and the peri-urban area known as Agricultural Park South Milan (Simeone and Cantú 2011; Cantú 2012).

As underlined by Van den Bosch (2010), scaling-up is less about scaling up products, services or users and more about scaling up perspectives, ways of thinking, routines, legislation, institutions, etc. Through scaling-up, socio-technical experiments can thus influence the way societal needs are fulfilled in a more sustainable direction (ibid.).

3.3.3 Socio-technical Experiments as Labs, Windows and Agents of Change

On the basis of the above considerations, socio-technical experiments can enhance the process of transitioning to sustainable radical innovations because they can simultaneously act as (Ceschin 2010, 2012, 2013a):

- **Labs**, to test, learn about and improve the innovation on multiple dimensions (technical, usability, regulative, political, economic, and socio-cultural) and in relation to different contexts (in order to learn about different PSS configurations in different settings);
- **Windows**, to raise interest in the innovation project and the related actors, disseminate results, build up synergies with existing similar projects/initiatives, and attract and enrol new actors (e.g. new users or potential partners); and
- **Agents of Change**, to influence contextual conditions in order to promote and quicken the societal embedding process. Experiments should be conceived to introduce and diffuse new ideas and knowledge to the community and stimulate various social groups (users, public institutions, companies, etc.) to change their perspectives, beliefs, and lifestyles. Experiments should in fact represent a stimulus to induce actors to change their behaviours and interpretative frames (for example, they can stimulate users to change their routines or push governmental institutions to implement proper policy measures to favour the PSS innovation).

3.3.4 Dynamics in Socio-technical Experiments (and Niches Formation)

It is important to underline that even if they can hold great promise, several radical innovations never move from a local experimental level to the mainstream level. So, why certain innovation journeys are a success and other a failure? What does it contribute to increase the chances of success? Many scholars explain success and failure through analysing the interaction between what has been named *three internal niche processes* (Kemp et al. 1998, 2001; Hoogma 2000; Hoogma et al. 2002; Raven 2005): the process of **building the social networks**, the process of **articulating expectations and visions**, and **the learning process**. In this section

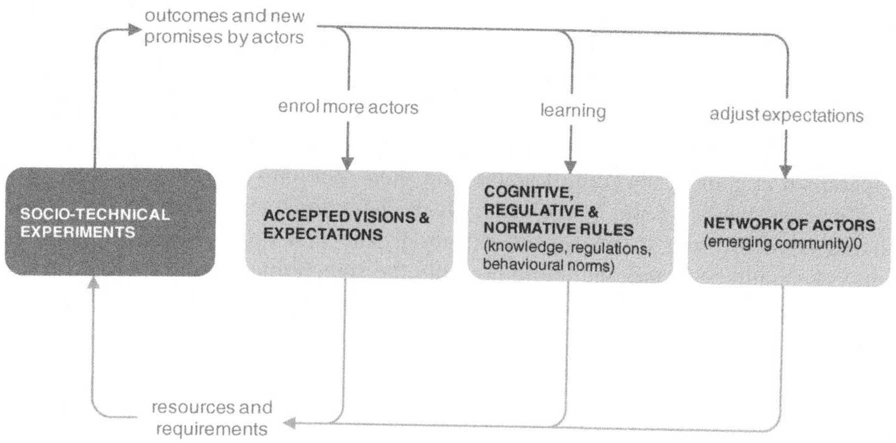

Fig. 3.3 The dynamics in socio-technical experiments. Adapted from Geels and Raven (2006)

we will briefly illustrate how these processes are interrelated, while in the subsequent ones we will elaborate on each single process.

The interactions between the three processes are illustrated in Fig. 3.3, and described by Geels and Raven (2006) as follows. Actors, embedded in networks, invest resources (money, people) in projects only if they have a shared positive expectation of a new innovation. This shared expectation provides a direction to the project and to the experiments. The outcomes of experiments produce learning processes, which are used to adjust previous expectations and enrol more actors to enlarge the social network. If outcomes of learning processes are positive (i.e. the initial expectation is validated) a new development cycle can be initiated (e.g. new experiments can be implemented). Expectation and shared rules remain stable, but become more detailed. Positive outcomes also facilitate the enrolment of new actors (potentially resulting in more available resources). If outcomes are below expectation, trust in the new innovation diminishes, expectations decline, and the social network may break up (potentially resulting in less available resources). In response to these negative outcomes actors tend to redirect the project (i.e. come up with new expectations and visions), changing the innovation journey course.

In short, the success of socio-technical experiments heavily depends on how the three processes are managed. The better these three processes are managed, the greater the possibility that experiments can develop into a market niche, influence and transform the existing regime or become a viable alternative to it. The following sections focus on these three processes.

Building up Social Networks In socio-technical experiments, and in the process of niches formation, the network composition plays a crucial role. This network is important to protect, support and foster radical innovations. Initially the network can be limited and fragile, meaning that few actors are involved in developing the innovation; if experimental activities meet the initial expectations the network can expand and become more stable. Raven (2005) points out that initially, actors' commitment to the project can be limited because they do not yet have many vested interests and withdrawal does not result in large losses. Moreover, in the beginning the network is usually unstable, meaning that the role of actors and their relations may be unclear. Only in the course of time, when more experience is gained, the role of actors and their relations becomes clearer.

However, the key issue is to clarify which characteristics the network of actors should have in order to be considered good. The first characteristic is that the network should be ***broad***, including not only the stakeholders more directly linked with the innovation (such as firms, partners, users, etc.) but also other relevant stakeholders from the science, policy and societal domains (e.g. research centres, governmental institutions, NGOs, special interest groups, etc.) (Raven 2005). In other terms the network should be heterogeneous and characterized by scientific, social, economic, politic and cultural linkages.

Hoogma (2000) argues that socio-technical experimentation and niche development require actors who are willing to invest in maintaining or expanding the niche for long time, even when short-term market value is absent (for example actors such as large firms). On the other hand Kemp et al. (1998) point out that these actors may also have vested interested in the incumbent regime, and therefore could be interested in slowing down the development process or orient it towards more incremental than radical innovations. Therefore, as suggested by Tushman and Anderson (1986) the involvement of actors that have no strong ties with the dominant regime is important (e.g. new firms or firms from other sectors), because they are more interested in introducing radical innovations. On the other hand these actors may have limited resources and therefore may not be able to maintain niche development for a long time.

On the same line of thought of Tushman and Anderson, Van de Poel (2000) argues that *outsiders* (with respect to the dominant regime) are needed in a network, because they do not share the current regime institutions and practices and therefore they may contribute in the development of innovations that deviate from that regime. In particular he suggests the involvement of three groups of outsiders: *outsiders firms* (because they can mobilise knowledge and financial and managerial resources to develop such alternative innovations), *scientists* (because they can introduce radically new designs, criteria, approaches and concepts), and *societal pressure groups* (because they have the potential to mobilise public opinion and insiders in the regime).

Weber et al. (1999) also state that *insiders* should be involved, and in particular that government support and protection can be of crucial importance: in the start-up phase in order to give experiments legitimacy and stability; and in the subsequent phases, in order to create widespread support for scaling up the new practices and institutions related to those experiments. In relation to the scaling up process, Van den Bosch and Rotmans (2008) also point out that key stakeholders to be involved are actors that have the power and willingness to directly influence the dominant culture, practices and institutions (such as *Ministries, policy makers and politicians*, etc.), and actors that (in)directly may influence the regime because they have an interest in embedding new sustainable practices in society (such as *NGOs, frontrunners in a sector or policy domain*, etc.).

The active involvement of users in all the stages of the innovation process, rather than considering them merely as sources of information, is of course key important (Von Hippel 1988; Hoogma and Schot 2001). Moreover, Hoogma (2000) also argues that the actors that are affected by the impact of the innovation, but that do not use the innovation themselves, should be involved (examples are neighbouring residents in the case of wind turbines, or environmental groups that represent general societal concerns).

In short, there is not a recipe to build a proper network of stakeholders, but in general it can be suggested to develop a broad network with a mix of insider and outsider actors. In general a broad stakeholder involvement is important because (Van de Kerkhof 2004):

1. it can legitimate decisions, as more stakeholders have been involved;
2. it can increase accountability, as the stakeholders involved have become co-responsible for the decision and related activities and action plans;
3. it can contribute to increase the richness of the process, due to the input a wider range of viewpoints, interests, information and expertise about the topic under consideration.

It must be stressed out that the involvement of a wide number of stakeholders in decision processes may also lead to management problems. For example stakeholders could defend their own interests rather than the network ones (Van de Kerkhof 2004). Moreover, management can be further complicated by tactical and strategic behaviour by stakeholders in decision-making processes (e.g. stakeholders can form alliances and have hidden agendas, or they can try to delegitimize other actors) (De Bruijn et al. 2002). Even for these reasons Van de Kerkhof and Wieczorek (2005) emphasise the need for a network manager (neutral to the content of the transition process, but well acquainted with the issues at stake and the related activities) capable to facilitate the process of selecting and managing the participants in the arena. Finally it is important to underline that the network should be managed dynamically because different stages of a societal embedding process require different network compositions (Weber et al. 1999). For instance some actors may have more relevance in the first phases and disappear in the following ones (e.g. a public administration can be involved only in the beginning in providing incentives and protection to the innovation).

The second network characteristic that has an influence in the outcomes of socio-technical experiments and niche development is the ***alignment of actors' activities***. This alignment refers to the degree to which actors' strategies, expectations, beliefs, practices and visions go in the same direction (Raven 2005). Different actors may have different expectations and visions about the purpose of an innovation, thus there is the possibility that their strategies may diverge considerably (ibid.). It is therefore important to avoid this misalignment. Rip (1995) suggests that the involvement of public authorities and other general interest actors, or relatively independent actors, can increase alignment in a network.

Moreover, alignment is also facilitated through regular interactions between the actors (Hoogma 2000), and by the collective development of visions and agendas.

Articulating Expectations and Visions As we have seen in the previous section, the process of experimenting and embedding radical innovations in the society requires the involvement of many actors from different domains. These actors may hold different visions of the future and different expectations about a particular innovation. Therefore, it becomes fundamental to manage the diversity of expectations, and their negotiation and alignment (Raven et al. 2008). The convergence of actors' expectations is important in order to give strategic orientation and legitimacy to the innovation development (Kemp et al. 1998; Raven 2005), and in order to attract new actors and resources (Raven 2005). This is particularly true in the first experimental stages when the innovation is still in early development and the network of actors is usually unstable and fragile.

The creation of a ***shared long term vision*** can contribute to align actors' expectations, and therefore to formulate agendas and action plans, and coordinate the strategies of the actors involved. Future visions, especially if generated in a participatory or collective process, can in fact be seen as shared constructions that may have the potential to guide actors and provide an orientation for joint action (Grin and Grunwald 2000; Quist et al. 2001). The power of a vision is that it can constitute at the same time a vehicle for interaction, communication, explanation and discussion among actors, and a vehicle for broader reflection on normative choices and effects (Dierkes et al. 1996). Moreover, a shared vision makes it possible to unite actors from different backgrounds (ibid.); and this is important because radical innovations require the linking of many different actors and networks, and the integration of knowledge from different fields. Smith et al. (2005) propose five functions of future visions for system innovations and transitions:

- *Mapping a 'possibility space'*: visions identify a set of plausible alternatives for socio-technical systems providing societal functions.
- *Providing a heuristic*: visions act as problem-defining tools by pointing to the technical, institutional and behavioural problems that need to be resolved.
- *Providing a stable frame for target setting and monitoring progress*: visions stabilise technical and other innovative activity by serving as a common reference point for actors collaborating on its realisations.
- *Providing a metaphor for building actor-networks*: visions specify relevant actors (including and excluding) acting as symbols that bind together communities of interest and of practices.
- *Providing a narrative for focusing capital and other resources*: visions become an emblem that is employed in the marshalling or resources from outside an incipient regime's core membership.

It has however to be underlined that expectations are not fixed (Van Lente 1993; Hoogma 2000; Raven 2005): actors can change their views and expectations, as result of the negotiation processes with other actors, but also in reaction to changes in the external environment. As a consequence the long term vision is continuously subjected to refinement and re-orientation. In other words this requires interpretative flexibility of the vision, enabling actors to align the vision with their own interests (Berkhout 2006).

Finally it has to be underlined that outcomes from experiments may change actor expectations (Hoogma 2000). If outcomes are positive, experiments can increase the robustness of expectations, because they may contribute to stabilise them. In other words expectations get more robust if they are increasingly based on tangible results from experiments.

Learning Processes Learning takes place when individuals assimilate new information and apply it to their subsequent actions (Hall 1993). It is therefore crucial in the process of inducing changes towards the adoption and diffusion of radical innovations, in which new basic belief, behaviours and rules are required (Van de Kerkhof and Wieczorek 2005). In fact, as underlined by Van den Bosch (2010) the learning process in experiments is aimed at contributing to develop new ways of thinking (culture), doing (practices) and organising (structure). An adequate learning process is considered key important in socio-technical experiments, because it enables adjustments of the innovation and increases chances for a successful diffusion (Raven et al. 2010).

Two main characteristics are considered important in a proper learning process. Firstly, the learning process should be ***broad*** (Weber et al. 1999; Hoogma et al. 2002), focusing on many dimensions of the problem: not only the technical and economic aspects of the innovation, but also the cultural (societal beliefs, values and habits), regulative (government policy and regulatory frameworks), and institutional (rules and norms) ones. In particular Hoogma et al. (2002) distinguish five main learning dimensions on which actors should focus in socio-technical experiments:

- *Technical development and infrastructure*: this includes learning about design specifications, required complementary technology/products/services and infrastructure.
- *User*: this includes learning about user characteristics, their requirements and the meanings they attach to a new innovation and the barriers to use they encounter.
- *Societal and environmental impact*: this entails learning about safety, energy and environmental aspects of a new innovation.
- *Industrial development*: this involves learning about the production and maintenance network needed to broaden dissemination.
- *Government policy and regulatory framework*: this involves learning about institutional structures and legislation, the government's role in the introduction process, and possible incentives to be provided by public authorities to stimulate adoption.

Secondly, learning should be ***reflexive*** (Hoogma and Schot 2001; Raven 2005; Kemp and Van den Bosch 2006), implying that both *first* and *second order learning* occur. First order learning concerns new insights on how to solve a given problem, without changes in problem definitions (e.g. learning about the effectiveness of a certain technology to achieve a specific goal). Second order learning concerns new insights at a higher level with regard to problem definitions, norms, values, goals and convictions of actors, and approaches on how to solve the problem (Brown et al. 2003). As underlined by Hall (1993) second order learning leads to a paradigm shift, a change in the problem definition, basic assumptions, norms, values and interpretive frames which govern the decision-making process of individuals, communities and organizations. Second order learning is crucial in transition processes because it represents a condition for implementing radical innovations. In fact, when learning results in changes in mental frameworks of actors and social groups, this increases the space for behavioural alternatives available to these actors and social groups

3.4 Introducing and Scaling Up Sustainable PSSs: Companies' Experiences and a Conceptual Framework

Previous sections have provided interesting insights on the dynamics that drive system innovations. In particular it emerged that (sequences of) socio-technical experiments can play a crucial role in triggering transitions, and that a proper management of these experiments can increase the possibility that experiments can develop into a niche, and influence and transform the existing regime (or become a viable alternative to it). In order to increase these chances it is emphasised the importance of: (1) *building broad social networks*; (2) *articulating and aligning actors expectation into a shared long-term vision;* and (3) *creating room for broad and reflexive learning processes.*

At this point the question is: ***to what extent these insights are relevant and valuable for the specificities of sustainable PSS innovations***?

In seeking to answer to this question the following text discusses which insights from transition studies are relevant for the specificities of sustainable PSS innovations. Building upon this discussion this section then presents a conceptual framework on how sustainable PSS innovations take place (and which are the influencing factors). The framework provides a structured overview of the factors considered important to increase the chances for a successful societal embedding process. A case studies analysis, investigating the innovation journeys made by six companies in introducing their PSS concepts in the market, is then used to validate and refine the framework.

3.4.1 Insights from Transition Studies and Their Relevance in Relation to Sustainable PSS Innovations

Socio-technical Experiments The *implementation of socio-technical experiments*, to test the technical, social, political and economic configuration of the innovation, and favour its societal embedding, could represent a potentially promising strategy for companies who want to shift towards a PSS oriented approach. The implementation of field tests and pilot projects, in order to learn about the technical and usability aspects of a solution, is a common strategy adopted by companies to test their innovations. However, the concept of socio-technical experiment does not refer to simple tests exclusively aimed at learning about the technical and usability aspects. It refers to a broader concept: a crucial emphasis is in fact given to the setting up of protected environments, to learning and improving the innovation on multiple dimensions (not only the technical, economic, market demand and usability aspects, but also the political, regulative, environmental, cultural and social ones), to stimulate and influence changes in the socio-technical context (in order to favour niches formation), and to the involvement of a wide range of actors. Thus, the adoption of this concept by PSS companies (and in general by PSS promoters) requires a completely new strategic attitude, focused not only on the PSS innovation (and how to test and improve it), but also on the socio-technical context (and how to influence it in order to create the most favourable conditions for the innovation).

Building Broad Social Networks In socio-technical experiments, and in the process of niches formation, the network composition plays a fundamental role. In fact the establishment and development of a broad socio-economic network, characterised by scientific, social, economic, politic and cultural linkages, is recognised a crucial process to protect, support and foster radical innovations. In sustainable PSS innovations the network of stakeholders that produces and delivers the solution represent a fundamental aspect because its configuration determines the sustainability potential of the PSS (in relation to this see Sect. 2.3). Therefore, for companies that want to adopt a PSS-oriented approach, network building is a crucial activity. However, transition scholars suggest to focus not only on the actors directly linked to the PSS solution (partners, suppliers, customers, etc.) but also on the actors that could provide support and protection to that solution (public administrations, governments, NGOs, etc.). As a consequence, a broader system approach should therefore be adopted by PSS promoters: they should focus not only on the PSS solution and its value chain, but also on the contextual conditions that may favour the societal embedding of the PSS innovation, and on the actors that could be involved to support this process.

Articulating and Aligning Actors' Expectation into a Shared Long-Term Vision As showed in the previous section, the creation of shared long term visions can help to align actors expectations and provide strategic orientation to the process of experimenting and embedding radical innovation in the society. The development of project visions, to be used as guides to formulate strategies and persuade potential partners and stakeholders, is a common activity done within companies (Marzec 2007). In relation to sustainable PSS innovations, the vision usually consists in a PSS idea or concept (Vezzoli 2007), which includes the general characteristics of the new business model, a sketch of the journey to achieve that vision, the motivations to develop it, and its potential benefits (ibid.). These visions are mainly used to communicate the PSS concept inside the company (e.g. to different departments) or outside (e.g. to project partners, customers, etc.). If a broader network of actors is involved in the process (as transition studies scholars suggest), the vision should be built in order to take in consideration also their perspectives and their potential roles.

Creating Room for Broad and Reflexive Learning Processes In relation to learning, broad and reflexive learning processes are considered crucial in the process of inducing changes towards radical innovations. As previously stated, the implementation of field tests and pilot projects, in order to learn about the technical and usability aspects of a solution, is a common strategy adopted by companies to test their innovations. However, this learning approach is usually limited, and does not include other important dimensions such for example the cultural, regulative and institutional ones. Therefore, the challenge for companies should be to broaden learning processes in order to include also these other dimensions. In relation to sustainable PSSs, this means that socio-technical experiments should be also used by companies to learn about the different contextual factors that could influence the innovation (e.g. learn about policy measures to promote the PSS, learn about the different barriers that could hinder the implementation of the innovation). The second challenge for companies is to induce learning (in particular second order learning) in other socio-economic actors (e.g. stimulate users in changing their habits, stimulate administrations in implementing appropriate policy measures, etc.). This means that companies should strategically try to influence context conditions in order to favour the adoption and diffusion of their PSS innovation.

3.4.2 A Conceptual Framework

Building upon the insights from transition studies (and their adaptation to the specific characteristics of sustainable PSS innovations), a conceptual framework is developed. It provides a comprehensive description on how eco-efficient PSS innovations take place and hypothesises the critical factors that have an influence on the process (see Fig. 3.4). In summary, the entry point of a sustainable PSS innovation is a project vision (1): a PSS idea or concept developed to overcome a societal/business challenge. This project vision provides a direction to the societal embedding process (2), in which a broad network of actors (3) experiments and learns how the project vision can be met. The societal embedding process is based on the implementation of small scale socio-technical experiments, the development and empowerment of a niche, and the scaling up of the PSS innovation (and its related new set of culture, practices and institutions) in the regime. The process is characterised by dynamic adaptation: what is learned by actors leads to a continuous and mutual adjustment of the transition path, the project vision and the actor network itself (4).

As illustrated in the previous sections, the setting-up of ***sequences of socio-technical experiments*** represents a promising strategy to support, hasten and orient the incubation, testing and maturation of radical innovations. In other words

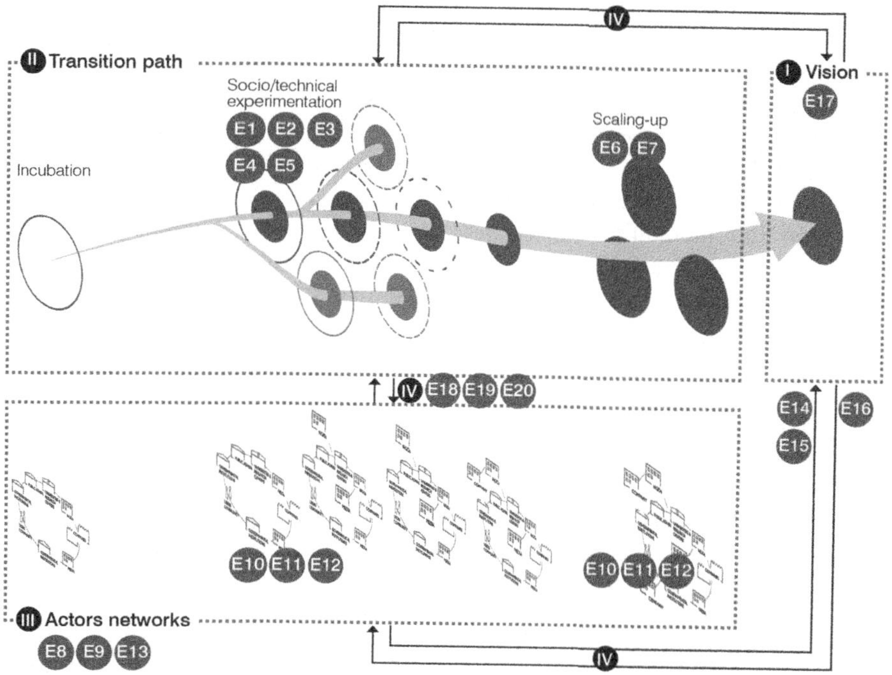

Fig. 3.4 Conceptual framework for the introduction and scaling up of sustainable PSS innovations. Numbers indicate the expected factors that can influence the process (see also Table 3.1). *Source* Ceschin (2012, 2013b)

the process of introducing and scaling-up radical innovations should be seen as a transition path, characterised by:

- an *incubation phase*, in which the conditions needed to start the societal embedding process are set up;
- a *socio-technical experimentation phase*, in which experiments are undertaken with the aim of learning and exploring how to improve the PSS innovation and how to contribute to its societal embedding;
- a *scaling-up phase*, in which the PSS innovation (and the related new practices, behaviours and institutions) increases its momentum and begins to influence the socio-technical regime (the initially unusual PSS innovation increasingly becomes part of the dominant way in which a societal satisfaction is fulfilled).

The literature review also showed that in this transition process a crucial role is played by the establishment and development of a ***proper network of actors***: a broad and dynamic network capable to protect, support and foster the innovation.

It is also fundamental to build-up a ***long-term vision***, shared among the actors involved in the project. In fact a shared project vision provides a direction to the societal embedding process and therefore a direction to stakeholders' actions.

Project visions are therefore important because they can be used as guides to formulate strategies, but also to attract and persuade new potential partners and stakeholders to join the project.

Finally, it is crucial the creation of room for ***broad learning*** (learning on many dimensions of the problem), and ***reflexive learning*** (learning resulting in changes in actors' reference framework, beliefs, behaviours, practices etc.).

Building upon these insights, Table 3.1 resumes the expected critical factors (based on the literature review) that can influence the process of societal embedding of eco-efficient PSS innovations. They are grouped in four clusters (societal embedding process, actor network, project vision, learning process) and positioned in Fig. 3.4.

3.4.3 Insights from Companies' Experience

In order to validate and refine the previously presented conceptual framework, a case studies analysis was undertaken.[7] The case studies analysis was structured in six steps [adaptation from Yin (1994)]:

1. *Conceptual framework development*: a first version of a conceptual framework was developed to provide a description of how eco-efficient PSS innovations take place and which factors influence that process. It represented the theory against which the cases were tested.
2. *Cases selection*: Cases are related to the innovation journeys made by six companies in introducing their PSS innovations[8] in the market. In particular two contrasting groups of cases can be identified: successful and unsuccessful ones. In this research a case is considered successful if the result of the innovation journey is (at least) the setting up of market niche in which the innovation is commercialized. Moreover, in order to obtain a broader picture of the phenomenon and facilitate the generalizability of results, heterogeneous cases were selected; in fact selected PSS innovations differ in terms of business sector and companies size. Table 3.2 provides a brief description of each case.
3. *Data collection methods*: In order to enhance the validation of the collected data, the case studies relied on triangulation (Yin 1994): multiple methods for collecting data were used to verify that all sources converged on the facts of a case. Within this research the sources of information are constituted by primary research data and/or secondary sources. Primary sources of data include questionnaires and semi-structured interviews with relevant companies' personnel (e.g. CEOs, directors, project managers, etc.). Secondary sources include

[7] This case studies research is deeply discussed in Ceschin (2013b).

[8] The PSS innovations selected are use-oriented PSS and result-oriented PSS. Product-oriented PSS have not been included in this study because they usually cannot be considered radical innovations and therefore they can be implemented using consolidated management strategies.

Table 3.1 List of the expected critical factors that can influence the process of societal embedding of sustainable PSS innovations

	Expected factors (in successful projects it is expected that…)	Source
Societal embedding process		
EF1	One or more socio-technical experiments are implemented	
EF2	Socio-technical experiments contribute to learn on many different dimensions (e.g. technical, user acceptance, political, regulative, cultural and social acceptance, etc.)	Kemp et al. (1998, 2001), Hoogma (2000), Hoogma et al. (2002)
EF3	Socio-technical experiments contribute to influence the socio-technical context in order to stimulate the societal embedding of the PSS innovation	Kemp et al. (1998, 2001), Hoogma (2000), Hoogma et al. (2002)
EF4	Socio-technical experiments are initially implemented in an environment protected from the mainstream market selection (financial protection)	Kemp et al. (1998); Weber et al. (1999); Hoogma et al. (2002)
EF5	Financial protection of socio-technical experiments is gradually dismantled during the scaling-up process, while institutional protection is reinforced and transferred into mainstream settings	Geels and Raven (2006), Rotmans and Loorbach (2006)
EF6	Scaling up is supported by repeating the experiment in a variety of contexts, and linking the experiment/s to other functions or domains	Schot and Geels (2008)
EF7	Scaling-up is favoured if there is an alignment of the niche with events and developments in the landscape	Geels (2005b)
Actor network		
EF8	The stakeholder network is broad (partners, suppliers, users, but also policy makers, governmental agencies, NGOs, research centres, media etc.)	Hoogma and Schot (2001), Raven (2005), Kemp and van den Bosch (2006)
EF9	The stakeholder network includes both insiders and outsider actors (with respect to the dominant regime)	Van de Poel (2000), Van den Bosch and Rotmans (2008)
EF10	Support and protection from governmental and public institutions could provide legitimacy and stability to the project	Weber et al. (1999)
EF11	The involvement of actors that have the power and willingness to directly influence the dominant regime (such as Ministries, agencies that develop protocols and standards, policy makers, politicians, directors, etc.) contributes to support the PSS experimentation and scaling-up	Van den Bosch and Rotmans (2008)
EF12	The involvement of actors that indirectly influence the regime because they have an interest in embedding sustainable practices in society (such as NGOs, sustainability ambassadors, frontrunners in a sector or policy domain, etc.) contributes to support the PSS experimentation and scaling-up	Van den Bosch and Rotmans (2008)

(continued)

Table 3.1 (continued)

Expected factors (in successful projects it is expected that…)	Source
EF13 The network composition is kept open to adjustment	Weber et al. (1999)
Project vision	
EF14 Actors expectations are aligned on a shared project vision and this contribute to develop a common and coordinated strategy	Kemp and Rotmans (2004)
EF15 The involvement of different actors in the development of the project vision (participatory approach) can contribute to build-up a shared vision	Kemp et al. (1998), Kemp and Rotmans (2004), Raven (2005)
EF16 The development of a clear and robust project vision can contribute to attract relevant actors and resources	Raven (2005)
EF17 The project vision is kept open to adjustment	Van Lente (1993), Hoogma (2000)
Learning processes	
EF18 Learning is broad: learning about the different dimensions of the innovation (e.g. institutional, technological, socio-cultural, environmental, economical)	Raven (2005)
EF19 Learning is reflexive: there is attention for questioning underlying assumptions such as social values, and the willingness to change course if the innovation does not match these assumptions	Raven (2005)
EF20 Reflexive learning is facilitated by the involvement of an heterogeneous set of different actors in the network	Hoogma and Schot (2001)

Expected factors are grouped in four main clusters. For each expectation it is reported the main source from which the expectation is drawn. *Source* (2012, 2013b)

Table 3.2 Overview of the cases analysed. For each case a brief description of the PSS innovation and its implementation and diffusion process is provided

PSS innovation
Clear Channel Outdoor, **SmartBike**, **USA** *B2C, Use-oriented PSS, Successful case* ***Description of the PSS***. SmartBike is a bike sharing system through which users can rent specifically designed bikes on a per minute basis. SmartBike is thought to be an alternative and integrative mean of transport, to be used for short trips supplemented by local public transport vehicles. About the organisation of the system, usually the local municipality (sometimes together with the local public transport company) covers the initial costs: the ones related to produce bikes and bike stations. Clear Channel Outdoor manages the service (and the related costs and revenues) ***Brief description of the implementation and diffusion process***. The solution was ideated in 1997. Initially the concept was proposed to several municipalities. After some rejections, in 1998 the concept was implemented as a small pilot project in the city of Rennes (France). This pilot was key important because gave the opportunity to test and improve the innovation (both from a technical and user acceptance point of view), and also to involve new actors (synergies were established with the local public transport company), and attract the interest of new users and other municipalities. In 1998 the system was scaled up in a full operational service and from 2001 replicated in 15 cities around the world
EGO, **Ecologico Guardaroba Organizzato**, **Italy** *B2C, Use-oriented PSS, Successful case* ***Description of the PSS***. EGO is a system for the shared use of dresses among a limited number of women. User, after the subscription, selects 14 clothes (from a sample book) to be inserted in the "shared wardrobe" (currently the "shared wardrobe" includes 120 models, grouped in 8 different styles). Once a week user selects and picks up 7 dresses, and at the same time brings back the dresses used during the previous week. User pays an annual registration fee plus a monthly subscription. EGO takes care of washing and maintenance. EGO not only manages the service but also designs the dresses and manages the manufacturing (outsourced to other Italian companies) ***Brief description of the implementation and diffusion process***. The PSS concept was ideated in 2003 by Vittoria Bono. From 2003 to 2008 the business idea was proposed, without having success, to several financial institutions to get funds. Without external financial support the concept was firstly implemented in 2008 as a small pilot project in the city of Brescia. This pilot was key strategic because gave the opportunity to learn about user preferences and improve the PSS offer. Thanks to the positive results obtained (and the support given by local environmental association and media), in 2009 a new and bigger point of sale was opened in Milan. In 2010 a franchising scheme was launched

(continued)

Table 3.2 (continued)

PSS innovation
Finnish Energy Service COmpanies (ESCOs), Finland
B2G/B2B, Use-oriented PSS, Successful case
Description of the PSS. In the ESCO model companies offer their customers a broad range of comprehensive energy solutions (e.g. designs and implementation of energy savings projects, energy conservation, energy infrastructure outsourcing, etc.). ESCOs gain their returns by receiving a share of the energy costs saved. If the project does not provide returns on the investment, the ESCO is often responsible to pay the difference. ESCOs are therefore economically incentivised in reducing the buildings energy consumption as much as possible
Brief description of the implementation and diffusion process. The ESCO model has been publicised in Finland for some years. However, the model in 2000 was not widely adopted. For this reason, in order to speed up ESCO activities, the Finnish Ministry for Trade and Industry (MTI) awarded investment subsidies to ESCO projects during 2001. In order to foster the diffusion of ESCO models, a network made up of ESCOs, municipalities, financial institutions, a governmental institution and a research centre, was established. Working groups and brainstorming sessions were organized in order to combine and match the needs and perspective of the different involved stakeholders, and to try to solve the implementation and diffusion barriers. As result pilot projects in collaboration with municipalities took place, and policy measures were adopted to favour the ESCO models. After the experimentation the number of ESCO projects in Finnish municipalities remained modest up to 2001 but rose considerably in the following years

(continued)

Table 3.2 (continued)

QURRENT, **Qurrent**, **The Netherlands**

B2G/B2B, *Use-oriented PSS*, *Successful case*

Description of the PSS. Qurrent enables people to produce and manage renewable energy by their own. Qurrent develops devices, software and services that enable the creation of small local energy networks (decentralised renewable energy systems). Within these networks users exchange energy to maximize the efficiency of the energy they produce. To do this Qurrent developed three core products: the Qbox (it measures all electricity production and consumption and makes it possible to share capacities with the neighbourhood), the Qmunity website (the place where Qurrent members go to analyse their energy consumption and production), the Qserver (where all measures by all members are stored); together they constitute the Local Energy Network

Brief description of the implementation and diffusion process. Igor Kluin (founder of Qurrent) presented his business idea in 2006 in an innovation contest, and won 250,000 €, which were used to develop the first generation of the Qbox. In 2007 Qurrent won another environmental prize, consisting of 500,000 € plus consulting and accounting help. Thanks to the prize Qurrent had the opportunity to set up seven pilot projects (currently running) to test the technical and usability aspects of its system, verify the acceptation by different kind of actors (private users, public administration, local energy suppliers, etc.), and learn about implementation and diffusion barriers. Since 2008 Qbox is on the market (available only in the Netherlands)

ARISTON, **PayXUse**, **Italy**

B2C, *Use-oriented PSS*, *Unsuccessful case*

Description of the PSS. PayXUse, developed by the Italian appliances manufacturer Ariston, is a PSS that provides to clients the access to washing machines (owned by Ariston). Payment is based on number of washes and includes: delivery of a washing machine at home (not owned by customers), electricity supply, maintenance, up-grading and end-of-life collection

Brief description of the implementation and diffusion process. The PSS concept was developed in 1999, in collaboration with ENEL (an Italian energy supplier). A first field test took place in 2000. In 2001, after solving the technical problems emerged during the field test, the PSS was launched in Lombardia and Marche (two Italian regions) in a limited amount of kits (around 500). At the end of 2001 the company decided to not launch the PayXUse in the market. The main reason that brought Ariston to this decision was the fact that the system was based on internally developed communication technologies and standards (not shared with the other main producers). Going on was therefore considered a risk because these standards were not shareable with the other main appliances producers (producers which at that time were planning to enter the market with a common communication standard and protocol)

(continued)

Table 3.2 (continued)

INTERFACE FLOORING SYSTEM, Evergreen Lease, USA
B2B/B2C, Result-oriented PSS, Unsuccessful case
Description of the PSS. As an alternative to a conventional purchase of carpet floors, clients "lease" the services (functionality, colour, design, aesthetics) of a modular carpet system without taking ownership or liability for maintenance and disposal of the products. A service package is offered inclusive of design layouts, product selection (choosing the right products for the right place), carpet installation, ongoing maintenance and ultimate removal for recycling. Evergreen was priced cheaper than the bought equivalent over the lifespan of a carpet
Brief description of the implementation and diffusion process. The PSS solution was ideated by Interface Flooring System in the first years of the '90 and marketed in 1995. Interface signed its first lease agreement in 1995 with the Southern California Gas Company. Many potential customers emerged enthusiastic for the carpet-leasing idea but unfortunately, despite the sales effort lead by the company, the majority of negotiations broke down. Only six lease agreements have been signed from 1995 to 2003. The reason of the missed success was mainly related to the financial accounting standards related to lease agreements, which made Evergreen Lease a capital lease instead of an operating lease, which is less favourable for customers. In addition some customers did not perceive the full cost of purchasing and maintaining carpets, which made difficult for them to understand the potential economic advantages of Evergreen Lease

Source Ceschin (2012, 2013b)
B2B Business to Business; *B2C* Business to Customers; *B2G* Business to Governments

companies' internal documents, scientific papers, and case studies made by other researchers.

4. *Data analysis and cases description*: collected data were analysed, selected and reduced, and used for the cases descriptions. A deductive approach was coupled with an inductive one. Firstly, the conceptual framework was used to examine if the factors identified in literature found a correspondence in the single cases (deductive reasoning). Secondly, the data collected from each case were used to hypothesise new relevant factors that were not identified in the literature analysis (inductive reasoning).
5. *Cross case analysis*: After describing the single cases and examining the interrelationship among the factors within each case, a comparison across the cases was undertaken. The aim was to verify if the identified factors were confirmed or rejected, and identify similarities and patterns among cases.
6. *Conceptual framework validation and refinement*: The results of the cross case analysis were used to validate and refine the conceptual framework built upon the literature review.

The following text, based on Ceschin (2013b), reports the results of the case studies research. For each of the four clusters it is discussed if the expected factors have been confirmed or not, and if new factors have emerged from the cases.

3.4.3.1 Societal Embedding Process and the Role of Socio-technical Experiments

The successful cases analysed in this research are all characterised by a commercialisation strategy based on the implementation of one or more socio-technical experiments. It has however, to be underlined that we are not dealing with simple experiments done inside a company's laboratory and exclusively aimed at improving the technical and/or usability aspects of the innovation, but with experiments undertaken in real settings involving a variety of actors and aimed at learning at many different dimensions (user preferences, political, regulative, cultural, social, etc.). For example in Finland, in order to foster the diffusion of Energy Service COmpanies (ESCOs), a network made up of ESCOs, municipalities, financial institutions, a governmental institution and a research centre was established; pilot projects were implemented in order to not only adapt the ESCO models to the municipalities' needs, but also to learn about the most effective financial and policy measures to support the PSS innovation (Ministry of Trade and Industry 2003; Kivisaari et al. 2004).

Experiments showed to be crucial to incubate and shape new PSS innovations (and the related socio-technical practices, habits and institutions), and also to stimulate and influence relevant actors to support and protect the innovations (and therefore favour their societal embedding). For example the bike sharing system ideated by Clear Channel Outdoor was firstly implemented as a small pilot project in the city of Rennes; the pilot was used to attract the interest of the local public transport company, involve it in discussions and negotiations, and finally develop synergies to stimulate public transport users to adopt the bike sharing system (Verrecchia S, 2009 "personal communication").

Moreover, in successful cases, experiments were also used to stimulate changes in actors' behaviours and habits. Continuing with the previous example, Clear Channel Outdoor states that the pilot was conceived as a way to show a new idea to the community, and stimulate potential users to get in touch with it and reflect about their mobility behaviours and routines (Verrecchia S, 2009 "personal communication").

In addition, another important element found in successful cases is the use of experiments as a means of communication to raise interest on the innovation project and the related actors. In other words experiments (and this is something that transition studies do not explicitly underline) were used as a sort of "windows" to disseminate the innovation results and facilitate the enrolment of new actors (e.g. new users, potential partners etc.). For example: when, in 1998, the bike sharing system in Rennes was improved and scaled up in a full operative service, Clear Channel Outdoor used it to demonstrate to other municipalities the potentialities of the PSS (e.g. municipalities were invited to visit the pilot); this was crucial in facilitating the gradual diffusion of the solution in other cities (Verrecchia S, 2009 "personal communication").

In short, successful cases are characterised by a strategic use of socio-technical experiments to: learn on many different dimensions (technical, usability, regulative, political, cultural and social acceptance, etc.); influence contextual conditions in order to favour the societal embedding process (stimulate changes in actors' behaviours and practices to protect and support the PSS innovations); communicate the PSS potentialities to attract and enrol new relevant actors. These characteristics cannot be found in the analysed unsuccessful cases, where experiments were mainly used to only verify and improve the technical aspects of the PSS. For example in the PayXUse case, the two pilot projects implemented in 2000 and 2001 were only focused on testing and improving technical issues (e.g. product requirements, data communication system), commercial issues (distribution channels), and usability issues (user interactions) (Aisa V, 2009 "personal communication").

Transition studies point out the importance of creating partially protected environments (financial and socio-institutional protections) where innovations can mature without the direct influence of dominant regimes (Schot and Geels 2008). Successful cases show that financial protection played a significant role in the first phases of experiments (setting up and implementation), while in the subsequent phases protection was gradually removed. On the other hand socio-institutional protection (new social relationships, routines, standards, etc.) was not dismantled, but rather reinforced and transferred into mainstream settings. For example, in the Finnish ESCOs case, the financing provided by the National Technology Agency of Finland (Tekes) was fundamental to set up initial workshops and roundtables, and implement pilot projects (Kivisaari et al. 2004). After the pilot project, institutional protection came from the Finnish Ministry of Trade and Industry, which inserted in its action plan for energy efficiency (2003–2006) recommendations on how to stimulate the ESCO models (i.e. new regulations, adoption of green public procurement) (Ministry of Trade and Industry 2003). In addition, institutional protection came also from the dissemination activities implemented by the Ministry to make Finnish municipalities aware of the ESCO models (Kivisaari et al. 2004; Ministry of Trade and Industry, Ministry of Transport and Communications, Ministry of agriculture and Forestry, Ministry of Finance, Ministry of the Environment 2007). However, as showed by the EGO example, successful implementation can take place even without protecting the first phases of the process. In this case PSS promoters tried, unsuccessfully, to get financial and institutional protection from the Italian Fashion Chamber and potential industrial partners; the missing protection resulted in a slowdown of the project development (Bono V, 2009 "personal communication"; Rovetta V, 2009 "personal communication"), but did not compromise its implementation.

Transition studies theorists argue that the implementation of sequences of experiments (and their repetition in different domains and contexts) is crucial to lead to the development and reinforcement of the niche (Raven 2005; Rotmans and Loorbach 2006). This is not fully confirmed by the successful cases analysed, which are characterised by the implementation of few experiments (1 in the EGO and SmartBike cases, and 3 in the ESCOs case). An exception is represented by the Qurrent case, in which 7 pilot projects were implemented to test their solutions in different settings (Qurrent 2009). Of course the implementation of a variety of experiments is important because can increase learning opportunities (e.g. learning about different designs in different contexts). On the other hand this strategy could be hindered by the limited financial resources that companies have for investing in experimental pilot projects [as in the cases of SmartBike and Ego (Verrecchia S, 2009 "personal communication"; Rovetta 2009)].

As underlined by transition studies, even if experiments in protected spaces are key important for transitions, scaling-up can take place if there is an alignment with events and developments in the landscape (Geels 2005b), and if the regime is sufficiently open to accept radical novelties (Rip and Kemp 1998). Successful cases show this correspondence. For example in the Finnish ESCO case, ESCO models were favoured by growing pressure from climate protection and growing energy prices (landscape development), which led to new energy taxes and stricter building regulations (regime development) (Kivisaari et al. 2004).

3.4.3.2 Actor Network

Successful cases are characterised by the involvement of a broad network, including actors from many different domains (e.g. users, governmental institutions, NGOs, special interest groups, research centres, etc.). A partial exception is represented by the EGO case, which shows a narrower network: in addition to suppliers and industrial partners, actors involved were potential users and (indirectly) environmental NGOs and media (Bono V, 2009 "personal communication").

More in details, successful cases show networks that combine outsiders and insiders actors (in relation to the regime). For example in the Finnish ESCO case, the network included: innovative companies and scientists from the Technical Research Centre of Finland on one hand (outsiders); municipalities, financial institutions and governmental institutions on the other (insiders). As underlined by van den Bosch and Rotmans (2008), this combination is important in order to couple the potentialities of outsiders in triggering innovations that deviate from the regime, and the potentialities of insiders in directly influencing regime culture, practices and institutions.

In particular successful cases demonstrate the crucial role that can be played by actors such as governments, governmental agencies, local administrations, etc. in directly creating a favourable environment for the PSS innovation. For example in the Finnish ESCO case, the Ministry of Trade and Industry and two governmental agencies (Tekes and Motiva) were decisive in funding the pilot projects, and also in adjusting regulations to facilitate the ESCO models adoption (Kivisaari et al. 2004).

An important role was also undertaken by actors such as environmental NGOs and societal pressure groups, because they contributed in disseminating the experiments results and mobilising public opinion. For example in the SmartBike case, the "publicity" provided by local NGOs was an important factor that brought to get exposure and attract potentially interested municipalities (Verrecchia S, 2009 "personal communication"). In addition, also media showed to be important in disseminating the PSS innovations. As declared by Rovetta (2009), the first customers of the EGO service decided to try it because they read an article on a newspaper or a specialised magazine.

In short successful cases show that companies adopted a strategic behaviours oriented at influencing the context in which the PSS should be introduced. In particular companies set up project networks trying to include those actors that, directly or indirectly, could have affected the regime (and thus create more favourable conditions for the adoption and further diffusion of the innovation).

The lack of a broad network is an important explanation of the relative failure of the PayXUse and Evergreen Lease cases. For example, in the PayXUse case, the main reason that brought the company to not introduce the solution in the market was the lack of a proper network capable to support the communication standard and protocols on which the solution was based (Aisa V, 2009 "personal communication"). Going on was therefore considered a risk because such standards and protocols were not shared with the other main appliances producers (ibid.). Building up relationships and agreements, with other producers and agencies that develop protocols and standard, could have brought to facilitate the adoption of Ariston's technology as a standard.

3.4.3.3 Project Vision

As illustrated in the previous section, the establishment of a broad network of actors is crucial in order to facilitate the societal embedding process of PSS innovations. In relation to this, transition studies point out that the development of proper project visions is fundamental in order to attract and enrol actors and give a strategic direction to the innovation (Kemp and Rotmans 2004; Kemp and Loorbach 2006; Raven 2005). Therefore it is key important for companies to be able to formulate clear visions (to make explicit their expectations), and be able to communicate those visions in an effective way. For example in the Qurrent case, the ability of the company founder in presenting the project vision (underlining all the potential economic and environmental benefits) was fundamental because it allowed to get the first funds (the Qurrent project idea was awarded in two important innovation contests), and because it facilitated the involvement of a variety of actors for the pilot projects implementation. The vision was structured in such a way in order to clearly illustrate how the solution would have worked and what (economic, environmental, etc.) advantages the implementation would have brought to the different stakeholders and the community. It resulted crucial for convincing funding bodies to invest on the idea, and for achieving a shared consensus (among the actors involved in the pilot projects) about the strategic direction to be followed.

It has however to be underlined that in successful cases the project vision was not kept fixed, but rather open to adjustments. For example in the SmartBike case, the initial project vision was modified once the public transport company in Rennes got involved in the project: the PSS solution was in fact refined in order to favour synergies between public transports and the use of SmartBike (Verrecchia S, 2009 "personal communication"). It is also important to not try to force consensus on a vision, but try to make explicit the different actors' expectations, and use these expectations as a basis for discussions and negotiations (Jolivet et al. 2003). This is exactly what was done in the Finnish ESCO case: before implementing the pilot projects, a 6 months incubation phase was set up in order to organise discussions, seminars and workshops (involving all the actors) with the aim to develop a shared ESCO model (Kivisaari et al. 2004).

The difficulties faced by the EGO in formulating and communicating its project vision in an effective way is an important explanation of the failure in establishing financial and industrial relations with other actors. As stated by Bono V (2009 "personal communication"), the Italian Chamber of Fashion, the financial institutions and the potential industrial partners contacted by the company, did not believe in the economic and environmental potential of the PSS innovation, and for this reason they decided to do not support the project.

3.4.3.4 Learning Processes

The analysed cases show that learning processes are strictly related to how experiments are designed and managed. The experiments that focused not only at exploring and testing the technical aspects, but also the ones related to usability, policy, regulations, social acceptance etc., easily brought to broad learning processes. For example in the SmartBike case, the pilot project brought to learn on technology (e.g. the docking station to lock and unlock bikes), user preferences (e.g. the service interactions to register and use bikes), regulations (e.g. how local administrations can support the solution), social acceptance (e.g. how to communicate the solution environmental benefits), strategic partnerships (e.g. how to build mutual synergies with the local public transport company) (Verrecchia S, 2009 "personal communication").

Successful cases not only show broad learning processes, but also reflexive ones. Reflexive learning is fundamental in order to break down actors' accepted assumptions and routine behaviours (Kemp and Van den Bosch 2006), and induce changes in culture, practices and institutions (Van den Bosch and Rotmans 2008; Brown et al. 2003; Brown and Vergragt 2008). For example in the SmartBike case, the first experiment implemented in Rennes induced reflexive learning in potential users (the novelty introduced by the pilot stimulated them to get in touch with the innovation and to potentially rethink about their mobility behaviours and routines), and in the local administration and public transport company (the pilot induced them to think about new alternative ways to plan mobility in city centres). As suggested by many authors, interactions among a heterogeneous set of actors (with different expectations, belief systems and interpretative frames) is a stimulus for reflexive learning (Hoogma and Schot 2001; Lynn et al. 2004; Brown and Vergragt 2008). This correlation between network heterogeneity and reflexive learning is confirmed in successful cases.

Another important issue is related to creating diversity among experiments, which is considered crucial because can led to learning about different designs in different use environments (Kemp and Loorbach 2003; Raven 2005; Van der Laak et al. 2007). This diversity can be found only in the Finnish ESCOs and Qurrent cases, were a variety of pilot projects were simultaneously implemented to learn about different contexts. For example in the Qurrent case, their decentralised renewable energy system was implemented in different settings: private families, housing corporations and office buildings (Qurrent 2009); this stimulated learning about different designs in different use environments. It has however to be underlined that the number of experiments and their degree of diversity depend on the available economic resources, and even if variety is considered important [e.g. to learn in different market contexts (Lynn et al. 2004)], companies could sometimes be financially limited in applying this strategy (as in the EGO case).

3.4.4 Conceptual Framework Refinement

Almost all the expected factors derived from the literature review were encountered in successful cases. One was partly confirmed (EF5). Two new factors emerged from the cases were added:

- NF1: Socio-technical experiments are used as 'windows' to disseminate the innovation results and facilitate the enrolment of new actors;

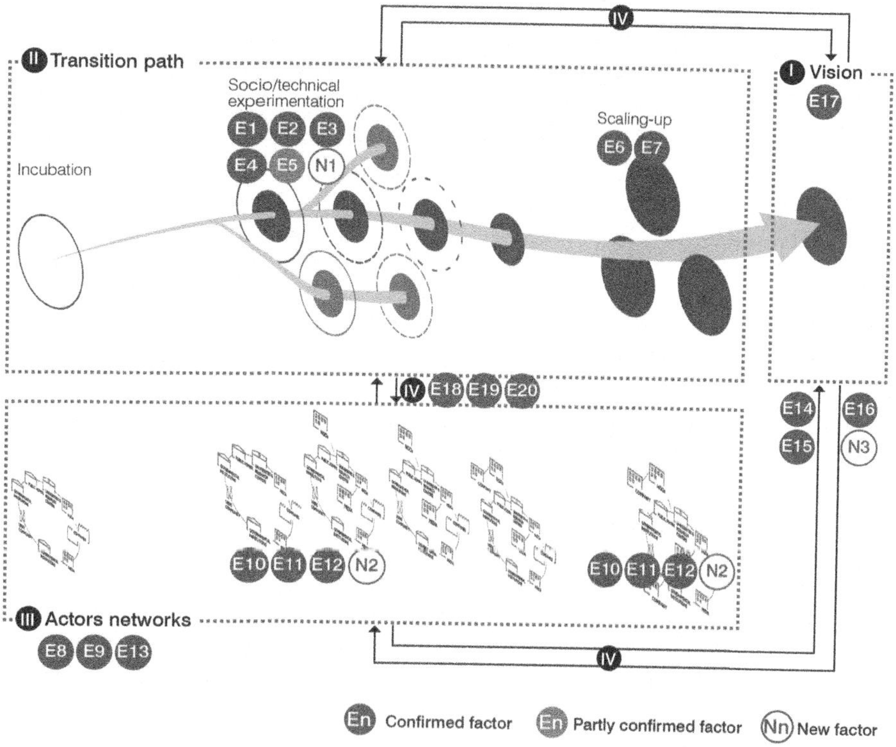

Fig. 3.5 Refined conceptual framework for the implementation and diffusion of sustainable PSSs. *Source* Ceschin (2012, 2013b)

- NF2: The stakeholder network is supported by actors that can mobilise public opinion (in particular media and NGOs).

Results are illustrated in Fig. 3.5.

It is important to underline that these factors are strictly interrelated. Socio-technical experiments require in fact the involvement of a broad network of actors. In order to involve actors, as well as to align their expectations, project visions are fundamental. The implementation of experiments stimulates learning processes, and what is learned by actors is continuously used to adjust the project vision, the network composition and the socio-technical experiments. Sequences of learning processes and experiment adjustments can allow the innovation to mature, and can give body to more stable socio-technical practices and institutions (capable to potentially influence the dominant regime).

It has to be emphasised that the list of factors provided by this research should be seen as general management indications, and not as a 'recipe for success'. Niche development and scaling up require in fact favourable conditions and circumstances [e.g. there should be enough pressure form the landscape, the regime should be sufficiently open to accept radical innovations, etc. (Rip and Kemp

1998)]. These conditions and circumstances may be not directly (or indirectly) influenced by companies. Therefore, the process from incubation to scaling-up becomes increasingly more uncertain and less manageable, and more influenced by project-external events and dynamics. However, the successful cases presented in this chapter show that the adoption of an experimental-, learning-, and network-based management approach, can increase chances of success (speed up and increase the possibilities to set up a market niche in which the innovation is commercialised).

3.5 Concluding Remarks

In Sect. 1.5 the first set of research questions (RQ1) was formulated: *How sustainable Product-Service System innovations take place? What are the dynamics and the factors that facilitate and obstacle the process of introduction and scaling-up? Is it possible to manage and orient this process? And if yes, how?*

In order to answer to these questions a desk research (literature review on transition studies) was combined with an empirical research (case studies). Based on the literature review a first version of conceptual framework was constructed. This conceptual framework provided a structured overview of the factors considered important to favour the societal embedding of sustainable PSS innovations. Case studies have been used to validate and refine the framework: expected factors have been confirmed or not, and new factors emerged from the cases have been added.

In short, the combination of literature review and case studies has provided significant insights and understanding on how the introduction and scaling up of sustainable PSSs takes place, and showed that an *experimental-*, *learning-*, and *network-based management approach* represents a promising strategy to increase the chances to successfully incubate, test, develop and bring to mature this kind of innovations.

Starting from these findings the next challenge is to understand the potential role that strategic design could play in this. In particular the next step is to answer to the following questions:

- Which role could strategic design play in supporting and orienting this kind of societal embedding process?
- Which design approach and tools can be used in practice to support strategic designers?

These issues are addressed in the next chapters of this research.

References

Berkhout F (2006) Normative expectations in systems innovation. Technol Anal Strateg Manage 18:299–311

Brown HS, Vergragt PJ (2008) Bounded socio-technical experiments as agents of systemic change: the case of a zero-energy residential building. Technol Forecast Soc Chang 75:107–130

Brown HS, Vergragt PJ, Green K, Berchicci L (2003) Learning for sustainability transition through bounded socio-technical experiments in personal mobility. Technol Anal Strateg Manage 13:298–315

Burns TR, Flam H (1987) The shaping of social organization: social rule system theory with applications. Sage Publications, London

Cantú D (2012) Ideas sharing lab. Community Centred Design for Multifunctional and Collaborative Food Services. PhD dissertation, Politecnico di Milano

Ceschin F (2010) How to facilitate the implementation and diffusion of sustainable product-service systems? Looking for synergies between strategic design and innovation sciences. In: Ceschin F, Vezzoli C, Zhang J (eds) Sustainability in design: now! Challenges and opportunities for design research, education and practice in the XXI century. Proceedings of the learning network on sustainability (LeNS) conference, vol. 1. Bangalore, 29 Sept–1 Oct 2010. Greenleaf Publishing, Sheffield, pp 440–454

Ceschin F (2012) The introduction and scaling up of sustainable product-service systems. A new role for strategic design for sustainability. PhD dissertation, Politecnico di Milano

Ceschin F (2013a) The societal embedding of sustainable product-service systems. Looking for synergies between strategic design and transition studies. In: Vezzoli C, Kohtala C, Srinivasan A, Xin L, Fusakul M, Sateesh D, Diehl JC (eds) Product-service system design for sustainability. Greenleaf Publishing, Sheffield (in press)

Ceschin F (2013b) Critical factors for managing the implementation and diffusion of eco-efficient product-service systems: insights from innovation sciences and companies' experiences'. J Cleaner Prod 45:74–88

De Bruijn JA, Heuvelhof EF, Veld RJ (2002) Procesmanagement: over procesontwerp en besluitvorming (in Dutch), 2 revised edn. Academic Service, Schoonhoven

Dierkes M, Hoffmann U, Marz L (1996) Visions of technology: social and institutional factors shaping the development of new technologies. Campus Verlag/St.Martin's Press, Frankfurt/New York

Elzen B, Geels FW, Green K (2004) Conclusion. Transition to sustainability: lessons learned and remaining challenges. In: Elzen B, Geels FW, Green K (eds) System innovation and the transition to sustainability. Edward Elgar, Cheltenham

Enerdata (2011) Global energy statistical yearbook 2011. Available on line: http://yearbook.enerdata.net/. Accessed 12 Aug 2011

Freeman C, Louça F (2001) As time goes by: from the industrial revolutions to the information revolution. Oxford University Press, Oxford

Geels FW (2002a) Technological transitions as evolutionary reconfiguration processes: a multilevel perspective and a case-study. Res Policy 31:1257–1274

Geels FW (2002b) Understanding the dynamics of technological transitions: a co-evolutionary and socio-technical analysis. Ph.D. Thesis, Twente University Press, Enschede

Geels FW (2004a) From sectoral systems of innovation to socio-technical systems: insights about dynamics and change from sociology and institutional theory. Res Policy 33:897–920

Geels FW (2004b) Understanding system innovations: a critical literature review and a conceptual synthesis. In: Elzen B, Geels FW, Green K (eds) System innovation and the transition to sustainability. Edward Elgar, Cheltenham

Geels FW (2005a) Processes and patterns in transitions and system innovations; refining the co-evolutionary multi-level perspective. Technol Forecast Soc Chang 72:681–696

Geels FW (2005b) Technological transitions and system innovations: a co-evolutionary and socio-technical analysis. Edward Elgar, Cheltenham

Geels FW, Elzen B, Green K (2004) General introduction: system innovation and transition to sustainability. In: Elzen B, Geels FW, Green K (eds) System innovation and the transition to sustainability. Edward Elgar, Cheltenham

Geels FW, Kemp R (2000) Transities vanuit sociotechnisch perspectief: achtergrondrapport voor het vierde nationaal milieubeleidsplan (NMP-4). Twente University and Merit, Twente and Maastricht

Geels FW, Raven RPJM (2006) Non-linearity and expectations in niche-development trajectories: ups and downs in Dutch biogas development (1973–2003). Technol Anal Strateg Manage 18:375–392

Grin J, Grunwald A (2000) Vision assessment: shaping technology in the 21st century towards a repertoire for. Springer, Berlin

Hall P (1993) Policy paradigms, social learning and the state. The case of economic policymaking in Britain. Comp Polit 25:275–296

Hegger DLT, van Vliet J, van Vliet BJM (2007) Niche management and its contribution to regime change: the case of innovation in sanitation. Technol Anal Strateg Manage 19:729–746

Hoogma R (2000) Exploiting technological niches: strategies for experimental introduction of electric vehicles. PhD Thesis, Twente University Press, Enschede

Hoogma R, Kemp R, Schot J, Truffer B (2002) Experimenting for sustainable transport: the approach of strategic niche management. Spon Press, London

Hoogma R, Schot JW (2001) How innovative are users? a critique of learning-by-doing and-using. In: Coombs R, Green K, Richards A, Walsh V (eds) Technology and the market. Edward Elgar, Cheltenham/Massachusetts, pp 216–233

Jégou F (2011) Social innovations and regional acupuncture towards sustainability. Chin J Des 214:56–61

Jolivet E, Laredo P, Shove E (2003) Managing breakthrough innovations: theoretical implications from and for the sociology of science and technology. Presentation for the ASEAT conference knowledge and economic and social change: new challenges to innovation studies, Manchester, 7–9 Apr 2003

Joore P (2010) New to improve: the mutual influence between new products and societal change processes. PhD thesis, Delft University of Technology

Kemp R (1994) Technology and the transition to environmental sustainability: the problem of technological regime shifts. Futures 26:1023–1046

Kemp R, Loorbach D (2003) Governance for sustainability through transition management. In: Proceedings of the international human dimensions programme (IHDP) conference, Montreal, Oct 2003

Kemp R, Loorbach D (2006) Transition management: a reflexive governance approach. In: Voss J, Kemp R, Bauknecht D (eds) Reflexive governance for sustainable development. Edward Elgar, Northhampton, pp 103–130

Kemp R, Rip A, Schot J (2001) Constructing transition paths through the management of niches. In: Garud R, Karnøe P (eds) Path dependence and creation. Lawrence Erlbaum, London

Kemp R, Rotmans J (2004) Managing the transition to sustainable mobility. In: Elzen B, Geels W, Green K (eds) System innovation and the transition to sustainability. Theory, evidence and policy. Edward Elgar Publishing, Cheltenham and Northampton

Kemp R, Schot J, Hoogma R (1998) Regime shifts to sustainability through processes of niche formation: the approach of strategic niche management. Technol Anal Strateg Manage 10:175–195

Kemp R, Van den Bosch S (2006) Transitie-experimenten. Praktijkexperimenten met de potentie om bij te dragen aan transities. Knowledge Centre for Sustainable System Innovations and Transitions (KCT), Delft

Kivisaari S, Lovio R, Vayrynen E (2004) Managing experiments for transition: examples of societal embedding in energy and health care sectors. In: Elzen B, Geels W, Green K (eds)

System innovation and the transition to sustainability. Theory, evidence and policy. Edward Elgar Publishing, Cheltenham and Northampton

Lynn GS, Morone JG, Paulson AS (2004) Marketing and discontinuous innovation: the probe and learn process. Calif Manage Rev 38:8–37

Marzec M (2007) Telling the corporate story: vision into action. J Bus Strateg 28:26–36

Meroni A (2008) Strategic design to take care of the territory. Networking creative communities to link people and places in a scenario of sustainable development. Keynote presented at the P&D Design 2008-8° Congresso Brasileiro de Pesquisa e Desenvolvimento em Design, Campus Santo Amaro, San Paolo, Brazil

Ministry of Trade and Industry (2003) Action plan for energy efficiency 2003–2006. Ad hoc committee reports 4

Ministry of Trade and Industry, Ministry of Transport and Communications, Ministry of agriculture and Forestry, Ministry of Finance, Ministry of the Environment (2007) Finland's national energy efficiency action plan (NEEAP 2008–2010)

Quist J (2007) Backcasting for a sustainable future. The impact after 10 years. PhD thesis, Delft University of Technology

Quist J, Knot M, Young W, Green K, Vergragt P (2001) Strategies towards sustainable households using stakeholder workshops and scenarios. Int J Sustain Dev 4:75–89

Quist J, Vergragt P (2000) System innovations towards sustainability using stakeholder workshops and scenarios. In: Proceedings of the POSTI conference on policy agendas for sustainable technological innovation, London, 1–3 Dec 2000

Qurrent (2009) Pilot projects available on line: www.qurrent.com/eng/projects.html. Accessed 29 Aug 2010

Raven RPJM (2004) Implementation of manure digestion and co-combustion in the Dutch electricity regime: a multi-level analysis of market implementation in the Netherlands. Energy Policy 32:29–39

Raven RPJM (2005) Strategic niche management for biomass. PhD dissertation, Technische Universiteit Eindhoven, Eindhoven

Raven RPJM (2006) Towards alternative trajectories? Reconfigurations in the Dutch electricity. Regime Res Policy 35:581–595

Raven RPJM, Heiskanen E, Lovio R, Hodson M, Brohmann B (2008) The contribution of local experiments and negotiation processes to field-level learning in emerging (niche) technologies. Bull Sci Technol Soc 28:464–477

Raven RPJM, van den Bosch S, Waterings R (2010) Transition and strategic niche management: towards a competence kit for practitioners. Int J Technol Manage 51:57–74

Rip A (1995) Introduction of new technologies: making use of recent insights from sociology and economics of technology. Technol Anal Strateg Manage 7:417–431

Rip A, Kemp R (1998) Technological change. In: Rayner S, Malone EL (eds) Human choice and climate change, vol 2. Batelle Press, Columbus, pp 327–399

Rotmans J, Kemp R, van Asselt M, Geels FW, Verbong G, Molendijk K (2000) Transities & transitiemanagement: De casus van een emissiearme energievoorziening (Transitions and transition management: the casus of a low emission energy supply). Maastricht University, Maastricht

Rotmans J, Loorbach D (2006) Transition management: reflexive steering of societal complexity through searching, learning and experimenting. In: van den Bergh JCJM, Bruinsma FR (eds) The transition to renewable energy: theory and practice. Edward Elger, Cheltenham

Rotmans J, Loorbach D (2010) Towards a better understanding of transitions and their governance: a systemic and reflexive approach. In: Grin J, Rotmans J, Schot J (eds) Transitions to sustainable development. New directions in the study of long term transformative change. Routledge, London

Schot JW (1998) The usefulness of evolutionary models for explaining innovation. The case of the Netherlands in the nineteenths century. History and Technology 14:173–200

Schot JW, Geels FW (2008) Strategic niche management and sustainable innovation journeys: theory, findings, research agenda, and policy. Technol Anal Strateg Manage 20:537–554

Schot JW, Hoogma R, Elzen B (1994) Strategies for shifting technological systems: the case of the automobile system. Futures 26:1060–1076

Schot JW, Lintsen HW, Rip A (1998) Methode en opzet van het onderzoek. In: Schot JW, Lintsen HW, Rip A, de la Bruhèze AA (eds) Techniek in Nederland in de twintigste eeuw: deel I. Walburg Pers, Eindhoven, pp 37–51

Scott WR (1995) Institutions and organizations. Sage Publications, Thousand Oaks

Simeone G, Cantú D (2011). Feeding Milan. Energies for change. A framework project for sustainable regional development based on for de-mediation and multi functionality as design strategies. Cumulus Shanghai conference 2010: young creators for better city and better life. Shanghai, 7–10 Sep 2011

Smith A, Stirling A, Berkhout F (2005) The governance of sustainable socio-technological transitions. Res Policy 34:1491–1510

Tuschman ML, Anderson P (1986) Technological discontinuities and organizational environments. Adm Sci Q 31:439–465

Tuschman ML, Romanelli E (1985) Organizational evolution: a metamorphosis model of convergence and reorientation. In: Cummings LL, Staw BW (eds) Research in organizational behaviour, vol 7. JAI Press, Greenwich, pp 171–222

Van de Kerkhof M (2004) Debating climate change: a study of stakeholder participation in an integrated assessment of long-term climate policy in the Netherlands. PhD thesis, Free University, Amsterdam. Lemma Publishers, Utrecht

Van de Kerkhof M, Wieczorek A (2005) Learning and stakeholder participation in transition processes towards sustainability: methodological considerations. Technol Forecast Soc Chang 72:733–747

Van de Poel I (2000) On the role of outsiders in technical development. Technol Anal Strateg Manage 12:383–397

Van den Bosch S (2010) Transition experiments. Exploring societal changes towards sustainability. PhD thesis, Erasmus Universiteit Rotterdam, Rotterdam

Van den Bosch S, Brezet JC, Vergragt PJ (2005) How to kick off system innovation: a rotterdam case study of the transition to a fuel cell transport system. J Cleaner Prod 13:1027–1035

Van den Bosch S, Rotmans J. (2008) Deepening, broadening, and scaling up: towards a conceptual framework for transition experiments. Knowledge Centre for Sustainable System Innovations and Transitions (KCT), Delft

Van den Bosch S, Taanman M (2006) How innovation impacts society. Patterns and mechanisms through which innovation projects contribute to transitions. In: Proceedings of the innovation pressure conference. Tampere, 15–17 Mar 2006

Van den Ende J, Kemp R (1999) Technological transformations in history: how the computer regime grew out of existing computing regimes. Res Policy 28:833–851

Van der Laak WWM, Raven RPJM, Verbong GPJ (2007) Strategic niche management for biofuels: analysing past experiments for developing new biofuel policies. Energy Policy 35:3213–3225

Van Lente H (1993) Chapter 6: construction by expectations. In: van Lente H (ed) Promising technology—the dynamics of expectations in technological developments. Twente University, Twente, pp 177–204

Verheul H, Vergragt PJ (1995) Social experiments in the development of environmental technology: a bottom-up perspective. Technol Anal Strateg Manage 7:315–326

Vezzoli C (2007) System design for sustainability. Theory, methods and tools for a sustainable "satisfaction-system" design. Maggioli Editore, Rimini

Von Hippel E (1988) The sources of innovation. Oxford University Press, New York

Walker W (2000) Entrapment in large technology systems: institutional commitments and power relations. Res Policy 29:833–846

Weber M, Hoogma R, Lane B, Schot J (1999) Experimenting with sustainable transport innovations: a workbook for strategic niche management. Universiteit Twente, Twente

Yin RK (1994) Case study research: design and methods, vol 5, 2nd edn., Applied Social Research Methods SeriesSage, Thousand Oaks

Chapter 4
Towards a New Way of Designing and Managing the Societal Embedding of Sustainable Product-Service System

Abstract This chapter investigates the potential contribution that a strategic design approach can make in triggering and orienting the societal embedding of sustainable PSSs. A new strategic design role emerges. A role in which the ideation and development of sustainable PSS concepts is coupled with the designing of appropriate transition paths to gradually incubate, introduce and diffuse these concepts. Starting from these considerations the chapter outlines and discusses the new design approach and capabilities required by strategic designers/project managers/consultants. The argumentation is accompanied by the discussion of an experimental design experience: the Cape Town Sustainable Mobility project.

Keywords Product-service system · Sustainable development · System innovation · Transition studies · Strategic design · Design for sustainability · Strategic niche management · Transition management · PSS implementation · Scaling up

4.1 Design and Radical Change for Sustainability

If we look to the evolution of Design for Sustainability (DfS) in the last decades, it clearly appears that this discipline has enlarged its field of action. In fact the focus has moved from single products only, to services and Product-Service Systems and, more recently, to social innovation and large-scale socio-technical changes.

Manzini and Vezzoli in 2003 emphasised the need for DfS to move from product thinking to system thinking (Manzini and Vezzoli 2003). In this respect they introduced the notion of *strategic design for sustainability* to suggest a design capable to:

- address sustainability operating on the integrated system of products, services and communication through which a company (or an institution, NGOs etc.) presents itself (Manzini 1999; Meroni 2008a; Vezzoli 2007);

F. Ceschin, *Sustainable Product-Service Systems*, PoliMI SpringerBriefs, DOI: 10.1007/978-3-319-03795-0_4, © The Author(s) 2014

- create a clear and comprehensible representation of the values and identity of a company and visualise possible futures to orient its strategy (Zurlo 1999; Borja de Mozota 1990);
- contribute to create relations between the stakeholders of a value constellation (Zurlo 1999), and act as facilitator to stimulate a strategic dialogue and co-design processes with them (Meroni 2008b).

A further shift took place when the DfS community started to investigate existing examples of bottom-up social innovations (Meroni 2007), and explore the potential role of design in this respect (Jégou and Manzini 2008). The emphasis has been on investigating the role of designer in triggering, supporting and orienting these community-based innovations through: the adoption of user-centred approaches to better understand problems and opportunities (ibid.; Mulgan 2009); facilitating participatory processes using visualisation techniques and co-design tools (Mulgan 2009); empowering individuals and communities enabling them to start and manage new bottom-up initiatives (Jégou and Manzini 2008). This is in line with the *transformative design* approach (Burns et al. 2006; Sangiorgi 2010), whose main aim is to *"leave behind the tools, skills and organisational capacity [to enable local actors] to continually responding, adapting and innovating"* (Burns et al. 2006).

The DfS research community made another step forward when recognised the necessity of moving towards the identification of strategies and approaches to implement, replicate and scale-up sustainable innovations. In other words, it was recognised that the challenge was no more only on proposing scenarios and concepts of sustainable changes but also on contributing to realise such changes (Vezzoli et al. 2008). Thus, researchers have started to explore the potential contribution of design in large-scale changes. The MEDEA institute at Malmö University proposed to use *Living Labs* to experiment, explore and support the scaling-up of grassroots social innovations. The concept of Living Lab refers in general to experimentation environments in which innovations are created in real life contexts by fostering collaboration between researchers, companies, end users and other relevant stakeholders (Ballon et al. 2005; Stålbröst 2008). The activities of Malmö Living Labs focus on supporting local actors in developing bottom-up social innovations by prototyping, testing and scaling-up solutions (Hillgren et al. 2011). They propose an open-ended approach, where prototyping is used to test and refine solutions, but also to create *"agonistic spaces"*, where a variety of stakeholders is involved in discussing and addressing dilemmas (ibid.).

In line with Malmö Living Labs, the Design Department of Politecnico di Milano proposed the concept of *"Enabling Experiment"*,[1] to refer to the implementation of favourable environments to enable local actors to take active role as

[1] The term was introduced by Ezio Manzini during his keynote speech *"To make things happen: Design as a catalyser of community engagement"* at the Design Pleasurable Product Interface 2011 conference (Milan, Italy).

co-creators in the development and proliferation of social innovations. An important contribution is that they acknowledged that large-scale changes require the implementation of a multiplicity of diverse and interacting experiments (Manzini and Rizzo 2011). In this respect Meroni (2008a) and Jegou (2011) refer to *"synergizing"* or *"acupunctural planning"*, the implementation of a set of synergic self-standing local experiments that aim to generate changes in large and complex systems (this is consistent with the *broadening* process mentioned in Sect. 3.3.2).[2]

The concepts of *Malmö Living Labs* and *Enabling Experiments* are consistent with the research results achieved by transition studies scholars. However, it must be stressed out that in these two approaches the characteristics of socio-technical experiments are not investigated and analysed in depth. Several issues are not addressed and remain unclear. In particular: How to set up protect spaces to shield the innovation in the first stages of the experimentation activities? What are the key dynamics in socio-technical experiments and what are the key factors influencing success and failure? What are the mechanisms through which a sequence of experience can lead to large-scale changes? What role can be played by different actors (users, policy-makers, local administrations, NGOs, consumer groups, industrial associations, research centres, etc.) in this process?

As shown in the previous chapter, transition studies have deeply explored the above issues and can therefore provide useful insights in order to better define and clarify the design approach to be adopted to conceive, implement and manage socio-technical experiments. On the other hand it must be underlined the important contribution that the previously mentioned researchers at Malmö University and Politecnico di Milano have made on analysing co-design approaches in the context of socio-technical experimentation (an aspect which is currently not deeply investigated by transition studies scholar).

From what it has been said above, it seems promising to create a bridge between the disciplines of strategic design for sustainability and transition studies.[3] The hypothesis is that the adoption and adaptation of principles and concepts from transition studies can enrich and advance the current debate on the role of design in large-scale sustainable changes.

Building upon this hypothesis, this chapter seeks to translate the insights and concepts from transition studies into something that can be used by strategic designers in their design practice. In particular the chapter presents an action research project, called *Cape Town sustainable mobility*, aimed at designing and

[2] As said in the previous chapter, an example of such kind of interconnected and synergetic experiments can be found in the *Feeding Milano* project, aimed at prototyping and implementing a network of "0 miles" food related services between the city of Milan and the peri-urban area known as Agricultural Park South Milan (Simeone and Cantù 2011; Cantù 2012).

[3] Also Gaziulusoy (2010) and Gaziulusoy et al. (2013) have explored the link between design and transition studies, but mainly at the product design level. They in fact developed a scenario method to link activities/decisions at the product development level in companies with the transformation which needs to take place at the societal level to achieve sustainability.

implementing a radical innovation: a sustainable mobility PSS for the disabled and elderly people in the suburban areas of Cape Town. The project was used to reflect on the design attitude adopted, and constantly develop insights on how to refine and make the design approach more applicable to practice. This was an iterative process because the design approach was implemented in a practical design experience and continuously developed, adjusted and refined during the whole design journey. Because the research had a non-linear (iterative) character, the order of the activities undertaken cannot be presented in a completely chronological way. Rather, for the sake of clarity, the choice is to firstly present the action research project and then the reflection on the design role and approach. In particular, building upon the project experience, the chapter discusses the potential role of design in triggering wider societal transformations by designing transition path (i.e. sequences of small scale socio-technical experiments); in this respect the chapter illustrates the *design attitude* and *skills* required by strategic designers to support and orient the societal embedding of sustainable PSSs, and discusses the *knowledge base* needed by designers to undertake this task.

4.2 An Experimental Design Experience: The Cape Town Sustainable Mobility Project

4.2.1 Project Background

The Cape Town Sustainable Mobility project involves, as main actors, *Shonaquip* (a small South African company producing wheelchairs and mobility equipment for disabled people), *Bicycle Empowerment Network* (BEN Bikes, a local association aimed at promoting sustainable mobility projects and initiatives), the *Cape Peninsula University of Technology* (CPUT), and *Politecnico di Milano* (Polimi). The aim of the project is to introduce and diffuse a sustainable mobility Product-Service System (PSS) for the disabled and elderly people in the suburban areas of Cape Town. In particular the system is expected to offer disabled and elderly people increased mobility services from their homes to the nearest public transport stops, or to local schools, hospitals, etc. Technically, the mobility system is designed around a solar, electric and human powered light vehicle.[4] This mobility system is especially conceived to create benefits in suburbs such as those in Cape Town, which are often characterized by substantial mobility problems due to a lack of high quality public transport services. The initial PSS concept was developed by Hazal Gumus for her master's degree thesis (Gumus 2009), conducted in collaboration between Polimi and CPUT. The thesis project raised the interest of Shonaquip and in July 2009, a process to socially embed the PSS innovation officially started.

[4] Prototyped by IPSIA "A. Ferrari" Maranello in 2006.

The initial assumption which drove the implementation strategy was that the setting-up of protected socio-technical experiments (with the characteristics described in the previous chapter) would have been promising to increase the chances to realise socio-technical changes. The activities undertaken in the project can be grouped in three main phases:

- *incubation*, aimed at setting up the conditions needed to start the societal embedding process;
- *socio-technical experimentation*, aimed at implementing the first socio-technical experiments, to learn and explore how to improve the PSS innovation and how to favour and support its societal embedding;
- and *scaling-up*, aimed at removing protection and transforming the experiments in a fully operative service.

At the time of the writing of this book the project consortium is in the between of the second and the third phases.

4.2.2 Incubation

The process began with the first formalization of the project vision, on the basis of the initial PSS concept. The aim was to translate the project idea into a set of visual artefacts to clearly and effectively communicate the PSS innovation's characteristics and its potential benefits to different types of actors. A set of visualisation tools was used to support this task.[5]

The next step was the development of a draft transition path, to identify the main steps between the present situation and a future situation with the PSS implemented. Actors involved in these first two steps were the research team (made up of academics and research students from *Polimi* and *CPUT*) and *Shonaquip*.

The following step was the identification of actors to be involved in strategic discussions. It was decided to firstly include a restricted group of actors (the ones considered crucial to start discussing and strengthening the PSS concept and the transition path) and later extend participation to a wider variety of actors such as the Cape Town municipality, the local public transport company, and local media. Actors initially involved were potential users, local citizens, technology experts from CPUT, and two local NGOs: *Disability Workshop Enterprise Development, DWDE* (active in providing job opportunities to disabled people) and *the Reconstructed Team* (an association aimed at reintegrating into society former drug addicts and criminals).

---

[5] Among them: (1) the *offering diagram*, to visualize which customer needs are addressed by the PSS; (2) the *interaction table*, to visualize how the PSS providers deliver the service and how the customers are to be satisfied; (3) the *system map*, to visualize the structure of the value chain; (4) the *sustainability diagram*, to visualize the environmental, socio-ethical and economic benefits.

A two-day workshop was organized in September 2009. The workshop began with the illustration of the project vision and the draft action plan; project promoters used the visual artefacts elaborated in the previous steps as a basis for the presentation. The first day focused on discussing and adjusting the project vision. In order to stimulate discussion, participants were asked to analyse the vision in relation to different socio-technical dimensions (technological, political, cultural, etc.) and identify conflicting issues. Participants were then asked to think about potential alternatives to solve the conflicting issues that had emerged. The collective discussion about the PSS concept and the context opportunities and barriers resulted in adjusting and refining the project vision at the end of the first day. The second day of the workshop focused on discussing the transition path, identifying: (1) steps and actions to be undertaken; (2) actors to be involved in the different steps; and (3) roles and tasks to be assigned to each actor. In short the workshop led to:

- *Adjust the PSS concept.* It was decided to also offer a transportation service for tourists within the city centre (in order to increase the sources of revenue). In relation to the vehicle the design requirements were specified.
- *Adjust the transition path.* In particular it was agreed that the next step would have been the implementation of a small-scale experiment in the Athlone district (focused only on the transportation of elderly people), to be later extended to other suburban areas of Cape Town.
- *Identify implementation barriers.* The main problem that emerged was the unavailability of financial resources to entirely finance the vehicles' production and the pilot implementation. Moreover another concern was related to the local availability of solar panels and lithium batteries. Finally it emerged that local regulations did not allow the use of human-powered vehicles for public mobility services.
- *Identify new actors to be involved.* It was recommended to establish connections with Cape Town municipality (and in particular the transport department) to solve the previously mentioned regulative issues and develop synergies with the public transport service. Moreover it was suggested to identify and involve an actor that could manage the tourist transportation service in the city centre.
- *Agree on the tasks to be assigned to each actor.* In particular it was agreed that Shonaquip would have managed the production of the vehicles (in collaboration with DWDE) and manage the service (in collaboration with the Reconstructed Team). CPUT would redesign the vehicle and contact the actors to be involved. Polimi would collaborate in the vehicle redesign and in seeking financial resources for the pilot project.

In summary, the result of this phase was the building up of a first network of actors and the development of a shared project vision and a first hypothesis of transition strategy. The involvement in this first phase of a broad variety of actors was crucial in order to allow the project consortium to focus on different dimensions of the problem (technical, economic, sustainability, usability, etc.). On the other hand, it emerged the difficulty to coordinate and manage discussions among a variety of actors, and the importance of having a network manager capable to manage controversies and conflicts within the network and establish bridges between different actors' expectations.

4.2.3 Socio-technical Experimentation

In second phase two socio-technical experiments were designed and implemented. The first experiment was implemented in the Athlone district (Bridgetown), in collaboration with the Reconstructed Team, and was aimed only at testing and improving the technical and usability aspects of the PSS innovation. In the first stage, before concluding the vehicle construction, an existing rickshaw was used to test the service of transporting the elderly in the neighbourhood, involving them in identifying critical issues and suggesting potential improvements (Figs. 4.1, 4.2). In the second stage, once the vehicle prototype was completed, a series of technical tests took place (Fig. 4.3).

After having settled the vehicle's technical problems and collected the first feedback on the service, the project consortium was ready to start the experimentation with users and the new vehicle. However, at this stage the *Reconstructed Team* decided to leave the consortium. It was an unexpected decision, even because of the positive response given by users during the service test. They explained the decision saying that due to other activities there were no personnel available to manage the complexity of this experimentation. On the other hand

Fig. 4.1 First socio-technical experiment: *photos* taken during the service test (July 2011). *Source* This photo was published in Ceschin (2012) but taken by Silvia Remotti

Fig. 4.2 First socio-technical experiment: *photos* taken during the service test (July 2011). *Source* This photo was published in Ceschin (2012) but taken by Silvia Remotti

Fig. 4.3 First socio-technical experiment: *photo* taken during technical test of the vehicle (August 2011). *Source* This photo was published in Ceschin (2012) but taken by Silvia Remotti

they confirmed the interest to implement in future, after the experimentation phase, a full operational service with a fleet of vehicle. The main lesson learned from this defection was that the presence of an internal catalyst, capable to promote and

Fig. 4.4 Second socio-technical experiment: testing the PSS (October 2011). *Source* This photo was published in Ceschin (2012) but taken by Emanuela Delfino

support the project inside each involved institution, is of key importance. Project promoters should have established stronger connections with key people in the *Reconstructed Team* in order to guarantee a more stable commitment.

At this stage the project consortium needed to find another actor willing to continue the experimentation and manage the implementation of a fully operative service. A contact was established with *BEN Bikes (Bicycle Empowerment Network)*. BEN Bikes is a local association aimed at promoting sustainable mobility projects and initiatives and providing job opportunities for low-income people. For this purpose they have several centres located in the suburban areas of Cape Town. The defection of the Reconstructed Team and the involvement of BEN Bikes led to the adjustment of the project vision. In particular BEN Bikes proposed to use their suburban hubs as operative centres to manage local mobility services and vehicle maintenance. For this reason the second socio-technical experiment was undertaken in collaboration with one of these hubs, and in particular the one placed in Lavender Hill suburban area. This second experiment was implemented in October 2011 and is still running. It was designed and organized in order to act as a *Lab, Window* and *Agent of Change*.

4.2.3.1 Experiment as Lab

The first aim of the experiment was to test and improve the PSS innovation (***experiment as Lab***). A service for the transportation of elderly, sick and disabled people from their home to any point of interest around the Lavender Hill community (such as to the hospital, church or the post office) was implemented and is currently running (Fig. 4.4). The main role of the local BEN Bikes centre is to manage the service as well as take care of vehicle maintenance. The experiment is currently used to:

- *test and improve the vehicle:* the role of BEN Bikes is to check the vehicle on a daily basis, in order to report the technical problems and identify potential solutions (in collaboration with Shonaquip, CPUT and Polimi);
- *test and improve the service:* the quality of the service is assessed using questionnaires and semi-structured interviews. Test users are asked to evaluate the service, identify critical aspects, but also to propose potential alternatives and improvements;
- *test and improve the PSS configuration:* verify the PSS configuration in terms of stakeholder value chain and business model and identify potential improvements to be implemented. Meetings involving project promoters are scheduled on a monthly basis to discuss these issues;
- *identify barriers:* the pilot is also used to identify potential implementation and diffusion barriers on multiple dimensions (e.g. socio-cultural and regulative). For this reason various actors (such as the local community, local institutions and NGOs) are involved to express their opinions, remarks and suggestions (regarding this see also experiment as a Window and Agent of Change). Of course most of the barriers were identified in the previous steps (during the incubation and the first experiment). However project promoters considered crucial to use the experiment to identify any further potential barriers.

It must be stressed that the approach adopted is aimed at favouring broad participation in the design choices. All the involved actors (from the potential users to the local community and the local institutions) are asked not only to evaluate and provide feedback on the project, but also to propose adjustments and alternatives.

4.2.3.2 Experiment as Window

The experiment is also designed to raise interest in the innovation project and attract and enrol new potential users and other relevant actors (***experiment as Window***). It represents a working prototype of how things could work, a conversation tool aimed at enhancing participation and enabling discussions with a larger audience of relevant socio-economic actors. With respect to this the BEN Bikes centre was conceived as a sort of 'open gallery' to allow visitors to see, touch and acquire information about the project (Fig. 4.5). Interested people can freely visit the centre and better understand the features of the project and its environmental, socio-ethical and economic benefits. Moreover demonstration visits are organized with specific actors (for example potential users but also potential future partners, local institutions, etc.). BEN Bikes personnel have been trained to be able to effectively describe the project and in particular to illustrate the potential advantages for different kinds of actors. This was considered particularly important by

Fig. 4.5 Second socio-technical experiment: interested people visiting the Lavender Hill BEN Bikes centre (October 2011). *Source* This photo was published in Ceschin (2012) but taken by Emanuela Delfino

project promoters because there was the need not only to disseminate information about the project but also to stimulate changes in actors' behaviour and routines (for example stimulate potential users to reflect on their mobility habits and consider the benefits that the solution could provide to them). This is strictly connected to the third function of the experiment: experiment as Agent of Change.

4.2.3.3 Experiment as Agent of Change

The experiment was also conceived to stimulate changes in actors' behaviour and habits and create favourable conditions for the introduction and diffusion of the PSS (***experiment as Agent of Change***). Therefore, in October 2011 an event for relevant actors was organized (Fig. 4.6). The aim of this event was to officially launch the experiment, illustrate the potential future developments, and discuss with invited actors how to support and create the conditions to accelerate the project. The event took place at the Lavender Hill BEN Bikes centre.

The actors invited to the event were:

- the *Cape Town municipality* (in particular the Transport department and the Environmental Resource Management department), because of their potential interest in the project and their direct influence on local transport regulation;
- *local actors* potentially interested in implementing specific mobility services based on the MULO vehicle: in particular local schools and the local clinic (Philiza Abafazi Bethu); and
- local media.

Fig. 4.6 Prof. Mugendi M'Rithaa (CPUT) welcoming the workshop participants. *Source* This photo was published in Ceschin (2012) but taken by Emanuela Delfino

The event was structured in four parts:

- a first part aimed at illustrating the project (economic, environmental and socio-ethical benefits) and presenting the socio-technical experiment;
- a second part aimed at illustrating the project future opportunities;
- a vehicle ride demonstration; and
- a workshop with participants to discuss the potential synergies that could be built to sustain and expedite the project.

The results of the event were positive. Firstly, local actors evaluated the project as valuable for local communities, because of its potential to bring about tangible economic, environmental and socio-ethical benefits. Secondly, one of the actors involved, the local clinic, stated their interest in implementing a service for the transportation of patients as soon as possible. Thirdly, the Transport department of Cape Town confirmed its interest in strengthening synergies between the PSS and the suburban bus lines. In addition the Transport department stated they would have planned meetings in their agenda to discuss the policy measures needed to support and foster the particular vehicle typology adopted in the PSS.

4.2.4 Main Intermediate Project Results and Next Steps

Starting from an initial PSS concept proposed by a small network of actors, the first project result is the building up of a broad network of actors and the alignment of their expectations towards the achievement of a shared vision. Currently, the main actors committed to the project are a company (Shonaquip), a NGO (BEN Bikes), a local institution (the Cape Town municipality), and two universities (Polimi and CPUT).

The second project result is the implementation of two socio-technical experiments in suburban areas of Cape Town:

- the first, in the Athlone district, tested a service for the transportation of elderly people and the technical aspects of the vehicle;
- the second, at Lavender Hill, is much more articulated and is still running. It is currently aimed at: testing and improving the whole PSS, raising interest in the project and enrolling new relevant actors, stimulating actors (such as potential users) to change their behaviour, routines and mental frameworks, and stimulating changes in the socio-technical context (such as changes in the regulative framework).

Because the second socio-technical experiment is still on-going, it is currently not possible to develop definitive conclusions. Nevertheless, it is possible to say that the whole journey strengthened the stakeholder network, served to refine and improve the PSS concept, and created important opportunities for future developments. In this regard it can be mentioned that:

- local actors located at Lavender Hill (the clinic and the school) stated their interest in implementing mobility services specifically dedicated to their needs;
- the 14 BEN Bikes centres, located in the Cape Town suburbs, could represent crucial hubs to replicate the experiment in other areas of the city;
- BEN Bikes is also interested in implementing a service for tourist transportation in the city centre;
- the Cape Town municipality is interested in creating synergies with the PSS and the public transport services (in particular in relation to the suburban bus lines).

4.3 A New Role for Strategic Design for Sustainability: Designing Transition Paths and Socio-technical Experiments

If we analyse the Cape Town Sustainable Mobility project from a design perspective we can see that design activities focused not only on ideating and developing a PSS concept, but also on understanding which strategies and development pathways are the most appropriate to support its introduction and scaling-up. A new design attitude is necessary to operate at such strategic level and, as a consequence, new strategic capabilities and knowledge base are required by designers.

4.3.1 A New Strategic Design Attitude

4.3.1.1 A Broader Design Scope

In the Cape Town Sustainable Mobility project, the first important consideration to be done is that *design* had a role not only in conceiving and developing the PSS innovation but also in supporting and catalysing the process of transitioning towards the implementation and scaling-up of the innovation. We can say that compared to the traditional *strategic design for sustainability* approach (Manzini and Vezzoli 2003), the approach adopted in the project was characterised by a **broader design scope**. In fact, in addition to the ideation and development of the PSS concept (the long term project vision), the focus has been in the *designing of a transition path* (Fig. 4.7).
In particular the design scope has been extended to:

- the *design of the sequence of steps* to gradually reinforce/improve the innovation and foster its societal embedding (incubation, socio-technical experimentation and scaling-up);
- and the *identification and involvement of the actors* that can support the societal embedding process in the various steps of the transition path.

In other words design focussed not only on generating a vision of how a mobility need could be met in an alternative and more sustainable way, but also on how to achieve that vision. In this transition path a crucial role is played by socio-technical experiments. As described in Sect. 4.1, some researchers in the design field (e.g. see Meroni 2008a; Jegou 2011; Manzini and Rizzo 2011; Hillgren et al. 2011) have recently started to explore the use of experiments to favour large scale changes. However, the main difference here is that experiments are not only used to test, improve and enhance radical innovations, but also to change the context in which the innovation should be introduced. In other words, following transition studies insights, experiments are conceived not only as *labs* and *windows*, but also as *agents of change* (i.e. to create the socio-cultural, institutional and organizational changes required to support the implementation and subsequent scaling up of the innovation).

4.3.1.2 A Multi-term Design Attitude

The first consequence of this design role is that design should simultaneously focus on different time frames. The Cape Town Sustainable Mobility experience showed that project actors adopted a **multi-term design attitude** (Fig. 4.8), because they simultaneously focussed on:

- the *project long-term goal (project vision):* the achievement of a future in which a the Cape Town Sustainable Mobility system is part of the usual way in which a particular mobility need is fulfilled; and

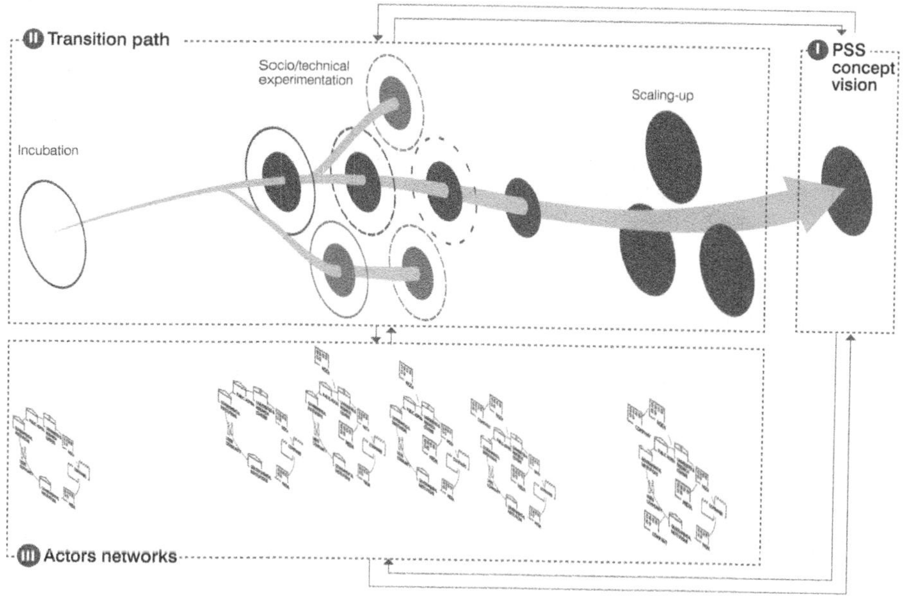

Fig. 4.7 A broader design scope. Design has a role not only in ideating and developing sustainable innovation concepts (i), but also in triggering and orienting transitioning processes: the designing of the sequence of phases and steps (ii), and identification of the actors to be involved along the whole process (iii). In these transition paths a key role is played by socio-technical experiments. *Source* Ceschin (2012)

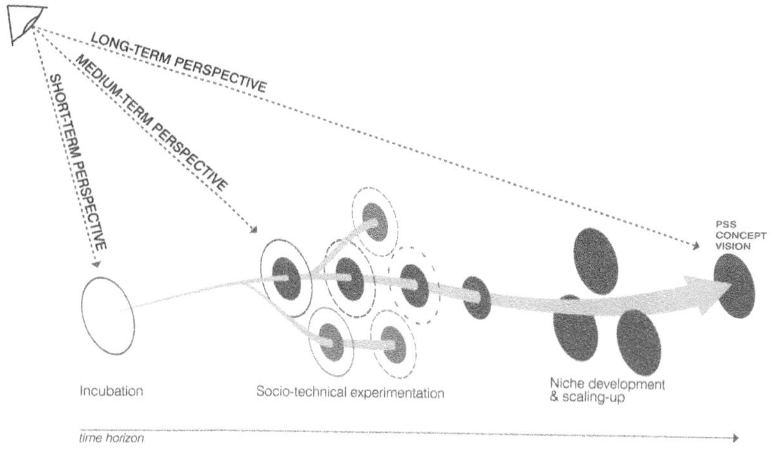

Fig. 4.8 The multi-term design attitude. The focus is simultaneously on different time perspectives. *Source* Ceschin (2012)

- the *short- and medium-term actions* to be undertaken in order to orient the innovation journey towards the achievement of the project vision: the incubation of the innovation and the implementation of two socio-technical experiments.

The project vision and the strategy to achieve the vision are not designed separately. Traditionally, the design of a solution is seen as a separate activity from the realisation of that solution. Here, there is not this dichotomy: the design of the project vision requires to be done simultaneously with the design of the transition path.

4.3.1.3 A Broader Strategic Design Attitude

In the notion of *strategic design for sustainability* (Manzini and Vezzoli 2003), the term 'strategic' refers to a design capable to address sustainability operating on the integrated system of products, services, communication and stakeholders value chain through which a company (or an institution, NGOs etc.) presents itself. However, designing experiments as agents of change requires the adoption of an even **broader strategic design attitude** (Fig. 4.9).

In fact the project actors should focus not only on the solution (the PSS innovation) but also on the technical, socio-cultural, institutional and organisational contextual conditions that might have favoured or hindered the societal

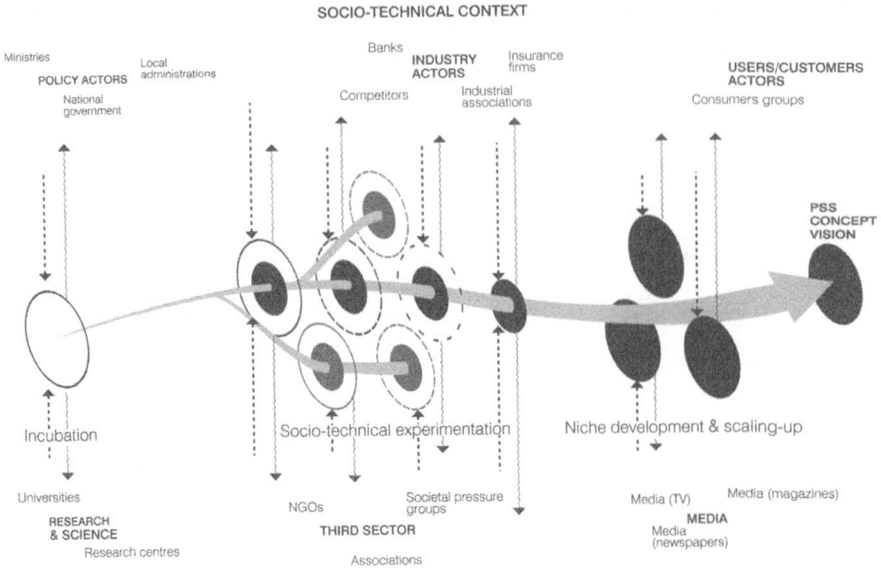

Fig. 4.9 A strategic attitude should be adopted in the designing and management of societal embedding processes. In this sense the transition path (and in particular the sequence of socio-technical experiments) is aimed at influencing changes in the socio-technical context in order to create favorable conditions for the introduction and scaling up of the innovation (*red arrows*). *Source* Ceschin (2012)

embedding process. In the Cape Town Sustainable Mobility experience the project consortium adopted a strategic behaviour because it tried to influence the socio-technical context, in order to create the most favourable conditions for the innovation. This was achieved by involving those actors that, directly or indirectly, could have affected regime practices and institutions and by stimulating changes in their behaviours, attitudes and practices. For example, one of the identified contextual barriers for the introduction of the PSS was related to the local road regulation. In order to solve this problem, PSS promoters involved in the project the municipality of Cape Town to stimulate the Transport Department to modify such regulation. Moreover the involvement of the Transport Department was also important to start developing proposals for the integration of the PSS concept with the local suburban public transport system.

Again, compared to existing approaches, the main contribution is that it is made explicit that experiments (and all the related activities, events, etc.) need to be deliberately designed in order to trigger changes in the socio-technical context in which they are inserted. And this through a strategic approach aimed at involving crucial socio-economic actors.

4.3.1.4 A Broader Co-design Approach

Researchers working on design and radical sustainable changes have always emphasised the importance of involving a broad range of actors in co-design processes. Meroni (2008a) speaks about *"community-centred approach"*, where the design focus is oriented towards the engagement of a community to activate local changes. Manzini and Rizzo (2011) and Hillgren et al. (2011) stress the importance of involving all the actors in co-designing within a process of continuous and open-ended experimentation. However, they do not make explicit the composition of the actors to be involved. In this respect transition studies scholars suggest that establishing and developing a broad and heterogeneous socio-economic network is crucial to protect, support and foster radical innovations (see Sect. 3.3.4). In particular it is crucial to involve outsiders and insiders actors (with respect to the dominant socio-technical regime): *outsiders* (e.g. outsider firms, scientists, societal pressure groups) are needed in a network because they do not share the current regime institutions and practices and therefore they may contribute in the development of innovations that deviate from that regime (Van de Poel 2000); *insiders* (e.g. policy makers, governmental institutions) should be involved because they can support and protect the innovation in the start-up phase (in order to give experiments legitimacy and stability) and in the subsequent phases (in order to create widespread support for scaling up the new practices and institutions related to those experiments) (Weber et al. 1999). In other words, it is required the involvement of a **broader network of actors** (Fig. 4.10). This is what the project actors tried to do during the Cape Town Sustainable Mobility design experience. In fact, they focussed not only on involving the actors that could have played a role in the value

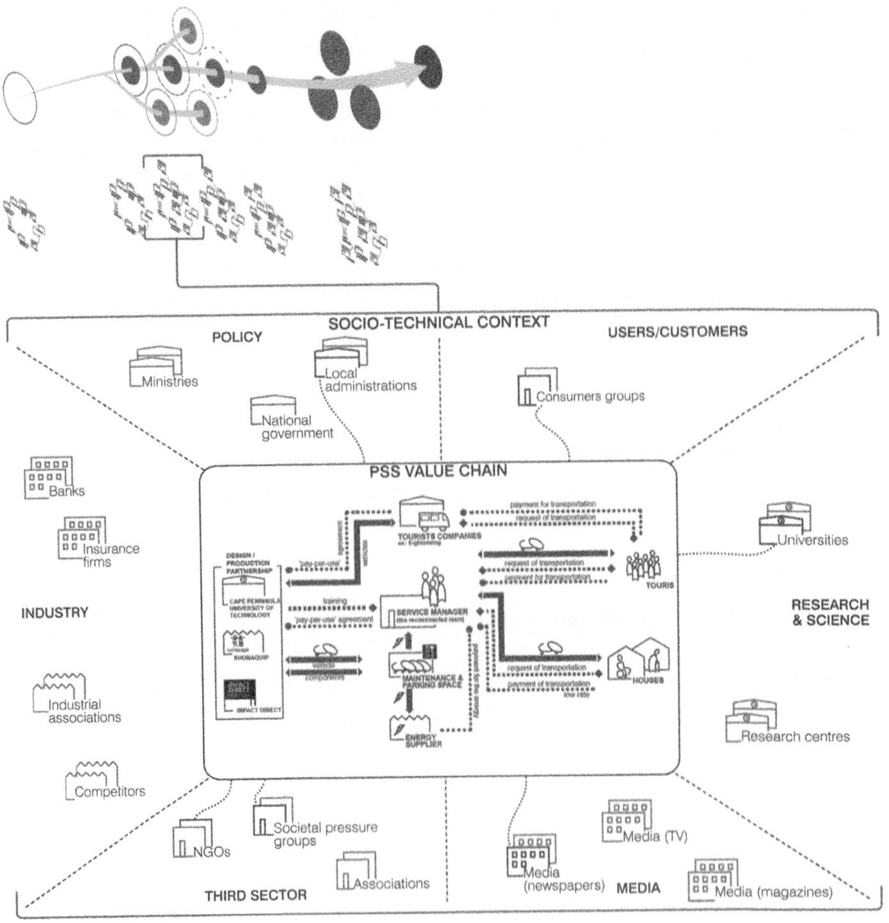

Fig. 4.10 The actors involved in the process of transitioning are not only the ones that are more directly linked with the innovation (value chain), but also the ones that could have an influence in the socio-technical context. *Source* Ceschin (2012)

chain (Shonaquip, BEN Bikes, suppliers, users, etc.), but also on other relevant actors belonging to the socio-technical context in which the PSS was being introduced (NGOs such as DWDE, the Cape Town municipality, local media, etc.). In other words they focussed on creating a broad network characterised by scientific, social, economic, politic and cultural linkages. Thus, when designing transition paths (and sequences of socio-technical experiments), it is crucial to involve those actors that can start a bottom-up process of change, but also those actors that can create favourable conditions to protect and support the innovation through top-down processes.

4.3.1.5 A Dynamic Design and Management Attitude

Finally, it is possible to say that the approach adopted in the project was characterised by a **dynamic design and management attitude** (Fig. 4.11).

The project vision was not a static outcome to be achieved; it was continuously adjusted as a result of changes in internal and contextual conditions and as a result of what was learnt by actors during the societal embedding process (in particular during socio-technical experiments). For example, the defection of the Reconstructed Team led to the involvement of BEN Bikes in the project network, which in turn led to adjusting the project vision (i.e. introducing the idea of developing mobility service based on BEN Bikes hubs and in synergy with the local public transport company). Adjustments in the project vision led of course to modifications in the transition strategy. Even the network of actors involved in the societal embedding process was dynamic: the composition, as well as the required tasks for each actor, continuously evolved in time as the different phases in the transition path required different network compositions. For example, the Cape Town municipality was involved only in the second phase of the societal embedding process, when institutional protection for the experiment was required.

In sum in the project it was therefore crucial to adopt a flexible and dynamic approach in order to manage these continuous re-directions and adjustments. An open-ended approach is required because of irreducible uncertainty, unpredictable events, changes in contextual conditions, and conflicting and alterable actors' expectations and views. This is in line with the idea of using experiments as *"vehicles able to raise questions as well highlight controversies and dilemmas"* (Hillgren et al. 2011), allowing conflicting voices to coexist.

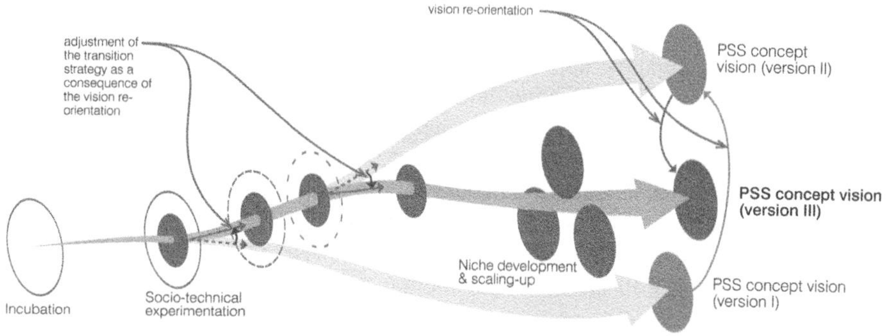

Fig. 4.11 A dynamic design and management attitude should be adopted. The project vision is not a static outcome to be achieved, and the transition strategy is not a fixed roadmap to be covered. Changes in internal and contextual factors, unpredictable events, learning process by project actors during the societal embedding process can lead to adjusting the project vision and, as a consequence, to re-orient the transition strategy. *Source* Ceschin (2012)

4.3.2 A New Set of Strategic Design Capabilities

Based on the discussion above, it emerges that new **strategic design skills** are required to design and manage the implementation and scaling-up of sustainable radical innovations:

- **Translating the project vision into a transition strategy**. Strategic designers should learn to translate a vision into the steps needed to support its implementation and scaling-up. In other words they must learn to design transition paths. Within these transition paths a crucial role is played by socio-technical experiments. Strategic designers should therefore learn to design these kinds of experiments and in particular to design sequences of experiments capable to act as *labs*, *windows* and *agents of change*.
- **Identifying and involving a broad variety of actors** to support the societal embedding process. Strategic designers should learn to identify the proper actors to be involved in the various phases of the process. Since the different phases of a transition path require different network compositions, strategic designers should be capable to design a dynamic network of actors: a network in which the composition, as well as the required tasks of each actor, continuously evolve in time. Moreover strategic designers should be capable of thinking not only about the actors that could be part of the value chain, but also about the actors that have the power and willingness to directly influence the dominant socio-technical regime. Strategic designers should thus be able to act as *networkers* (capable of establishing bridges and links between different actors) and as *negotiator* (capable of managing controversies and conflicts within the network).
- **Facilitating the building up of a shared project vision and transition path**. Strategic designers must learn to facilitate the strategic conversation between the actors involved, in order to develop (and adapt in time) a shared project vision and transition path. Strategic designers should therefore be able to facilitate a participatory approach, involving a variety of stakeholders in discussing, negotiating, co-creating and developing alternatives.[6] It is therefore crucial for strategic designers to be able to organize the complexity of the information that must be exchanged and support effective communication activities among stakeholders; encourage and stimulate the various actors in taking part in strategic conversations; ensuring mutual understanding; and manage the diversity of their expectations as well as their negotiation and alignment. These skills are thus fundamental: being a *communicator* (capable of effectively illustrating complex information such as project visions and action plans) and a *facilitator* (capable of activating co-design processes and facilitating the convergence of actors' expectations).

---

[6] In this respect see for example the use of communication strategies and tools in the Funny Dunny project (Mellick Lopes et al. 2011).

- **Managing the dynamic adaptation of the societal embedding process.** Strategic designers should learn to manage the continuous adaptation and evolution of the project vision, the transition path and the actor network. The societal embedding of an innovation should therefore be managed not as a project with a fixed result, but rather as an open search and learning process. Design, development, experimentation and implementation should be carried out simultaneously and in continuous interaction.

4.3.3 A New Knowledge Base

It must be stressed out that these new required capabilities and skills must be grounded in a proper knowledge base. In this sense the knowledge base required by designers to work at such strategic level must be enriched with new concepts and notions. In particular:

- First, strategic designers must know how radical innovations (and in particular sustainable PSS innovations) take place, and which are the main related dynamics and mechanisms. It is of key importance for them to be aware that these innovations are highly complex and uncertain, because they require multi-dimensional changes, at both the technical and socio-cultural level. It is also fundamental to know the different factors that could hinder or favour the process of introduction and scaling up of sustainable PSS innovations.
- Second, strategic designers must know the role that socio-technical experiments and niches can play in the process of introducing and scaling up radical innovations. They must be aware that the implementation of sequences of socio-technical experiments can favour, support and hasten the societal embedding of sustainable PSSs. In particular they should be aware of the mechanisms through which experiments can contribute to transitions (deepening, broadening and scaling up), and the role that a proper social network, a proper articulation of expectations and visions, and a proper learning process can play in successfully managing socio-technical experiments.

4.4 Design Process

The previous sections illustrated the design attitude and skills required by strategic designers to support and orient the societal embedding of sustainable PSSs. However, in order to operatively equip strategic designers, there is the need for a practical *'how to do it' design process*, and associated guidelines and tools. This section focuses on presenting this *design process* and providing an overview of the *tools* that can be used along the design process.

The design process proposed in this section is built upon the experience in the *Cape Town Sustainable Mobility* project. It has been developed in order to be used by strategic designers, project managers, and consultants to support and guide a company (or a small network of actors) in managing and enhancing the societal embedding process of a sustainable PSS innovation. In particular it is thought for practitioners who are already familiar with the concept of sustainable PSS innovation and know how to design sustainable PSS concepts (in fact the design process is thought to be used after the designing of the first version of the PSS concept). These practitioners can work in various contexts and organisations: consulting agencies, companies, research institutions, governmental departments, etc.

In particular the design process aims to facilitate the adoption of an *experimental-, learning-, and network-based design and management attitude*, seeking to increase the chance of successfully introducing and diffusing sustainable PSSs. It is thought to enable strategic designers and project managers to act as facilitators and mediators in the whole process, stimulating and encouraging the company and in general all the involved actors in: (1) formulating a shared project vision; (2) focusing on the contextual conditions that may favour or hinder the societal embedding of the project vision; (3) developing an action plan and adopting a strategic attitude oriented to influence the socio-technical context in order to create the most favourable conditions for the innovation; (4) adopting a reflective attitude to continuously reflect on and learn from the activities undertaken. The entire list of criteria used for the development of the design process is presented in Table 4.1.

The development of the design process builds upon the theoretical reflection on four instrumental approaches in the field of system innovations and transitions: *Strategic Niche Management* (*SNM*) (Kemp et al. 1998; Weber et al. 1999), *Conceptual Niche Management* (*CNM*) (Hegger et al. 2007), *Transition Management* (*TM*) (Rotmans et al. 2000; Loorbach 2002, 2007; Kemp and Loorbach 2003) and *ESTEEM* (Raven et al. 2009). The design process combines and adapts concepts and insights from these four approaches with a strategic design perspective.

4.4.1 Phases

The literature review from transition studies and the case study analysis (Chap. 3) showed that the setting-up of sequences of socio-technical experiments represents a crucial step to support and speed-up the incubation, testing and maturation of sustainable PSSs, and potentially facilitate their scaling up. However these experiments do not develop in a vacuum; rather it is necessary to set up proper conditions to start the societal embedding process. Furthermore these experiments do not automatically guarantee that the innovation will be scaled up in future; rather it is necessary to adopt appropriate strategies to increase the chances for a successful scaling up.

Thus, as proposed in the conceptual framework, we can distinguish three main phases in the process of introducing and scaling-up radical innovations:

Table 4.1 Criteria adopted to develop the design process

Criteria
1. The design process should enable strategic designers, project managers, and consultants to support and guide a company (or a small network of actors) in managing and enhancing the societal embedding process of a sustainable PSS innovation
2. The design process should stimulate project participants to adopt a broad strategic approach, and focus not only on the PSS innovation but also on the contextual conditions that may favour or hinder the societal embedding of the PSS itself
3. The design process should enable strategic designers and project participants to adopt a 'multi-term perspective', and simultaneously focus on the long-term goals and the short- and medium-term actions to be undertaken
4. The design process should equip strategic designers and project participants to formalise project visions
5. The design process should equip strategic designers and project participants to translate the project vision into a transition strategy (action plan)
6. The design process should enable strategic designers to design pathways of socio-technical experiments capable to act as labs, windows and agents of change
7. The design process should help to identify the proper actors to be involved in the various phases of the societal embedding process
8. The design process should facilitate a participatory approach
9. The design process should stimulate strategic designers and project participants to adopt a reflective attitude, bringing them to continuously reflect on and learn from the activities undertaken, and in case reconsider underlying assumptions
10. The design process should enable strategic designers and project participants to dynamically manage the societal embedding process (and continuously adjust project vision, action plan and network of actors)

- an *incubation phase*, in which the conditions needed to start the societal embedding process are set up;
- a *socio-technical experimentation phase*, in which experiments are undertaken with the aim of learning and exploring how to improve the PSS innovation and how to contribute to its societal embedding;
- a *scaling-up phase*, in which the PSS innovation (and the related new practices, behaviours and institutions) increases its momentum and begins to influence the socio-technical regime (the initially unusual PSS innovation increasingly becomes part of the dominant way in which a societal satisfaction is fulfilled).

These three phases constitute the backbone of the design process (see Fig. 4.12). Below a brief description of each phase is provided.

Incubation The incubation phase is concerned with the setting up of the conditions needed to start the societal embedding process. The starting point is a sustainable PSS concept vision developed by one or more actors (e.g. a company, a public institution, or a network of heterogeneous actors) as an answer to a societal/ business challenge. Actors potentially interested in the concept, as well as actors needed to provide protection and support to the innovation, are identified and involved. A first network of actors is therefore built around the PSS concept. Discussions and negotiations take place in order to achieve a common consensus on the PSS concept as well as on the potential strategies to socially embed the

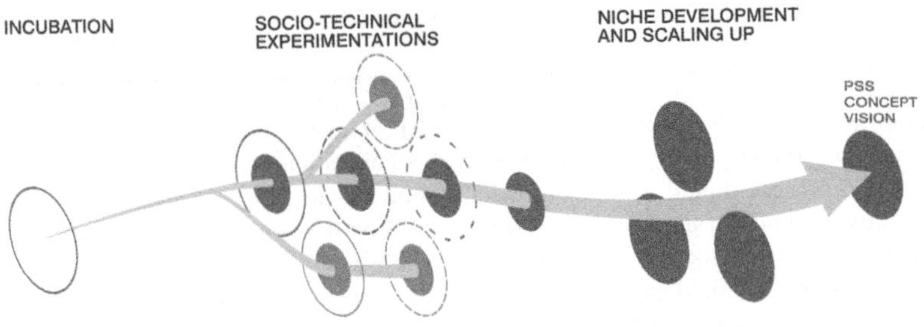

Fig. 4.12 The three phase of the design process. *Source* Ceschin (2012)

concept. The result of this phase is the setting-up of a network of actors who share the aim of achieving a long-term vision through an agreed transition strategy (action plan).

Socio-technical Experimentation In this phase socio-technical experiments are designed and implemented with the aim of learning and exploring how to improve the PSS innovation and contribute to its societal embedding. These experiments are undertaken at a local scale, in an environment protected from market competition (alternative selection environment), and involve a broad network of actors (e.g. a consortium including multiple companies, potential users, a public authority, NGOs, etc.). This protected space enables actors to explore and learn about local shifts in culture (new ways of thinking, value, and perspective), practices (new ways of doing, habits and routine) and institutions (sets of rules and procedures that guide the interactions and behaviours of actors). Particular importance is given to the conception and implementation of experiments capable to act as: *Labs*, to test and improve the PSS innovation on multiple dimensions and in relation to different contexts; *Windows*, to raise interest on the innovation project and the related actors, disseminate results and attract and enrol new actors; and *Agents of change*, to influence contextual conditions and favour the societal embedding process.

Scaling up Continuous experimentations and their linking with other existing projects and initiatives can lead to the development and empowerment of the innovation. At this stage the aim is to increase momentum of the PSS innovation (and the related new practices, behaviours and institutions), and start to have an influence on the socio-technical regime in terms of expectations, visions, knowledge, rules, etc. It is a process that can lead the PSS to increasingly become part of the mainstream way in which a societal satisfaction is fulfilled. Key issues to be taken in consideration are: (1) to replicate the PSS innovation in other contexts and create synergies with similar projects and initiatives at a broader level; (2) to disseminate information/project results and stimulate media attention at a national level; (3) to stimulate actors at a strategic level to influence the socio-technical context in order to create the most favourable conditions for the scaling-up of the PSS innovation.

4.4.2 Activity Clusters

The literature review from transition studies, the case studies analysis and the action research project showed that there are crucial activities to be managed along the whole societal embedding process. Strategic Niche Management (SNM) underlines the importance to properly manage the *"three internal niche processes"* (Kemp et al. 1998, 2001; Hoogma 2000; Hoogma et al. 2002; Raven 2005): the process of *building the social network*, the process of *articulating expectations and visions*, and the *learning process*. Transition Management (TM) puts forward a transition cycle characterised by four sets of activities (Loorbach 2002, 2007; Loorbach and Rotmans 2006): (1) *establishing and developing a transition arena;* (2) *developing sustainability visions and transition agendas,* (3) *initiation and execution of transition experiments and projects;* and (4) *monitoring, evaluating and learning.* In sum, the main activities to be managed during the societal embedding process can be grouped in five clusters:

- **vision building and expectation shaping** (*WHAT* the project network wants to achieve): activities related to building-up and formalising a project vision shared among the actors and social groups involved in the project;
- **action plan development** (*HOW* the vision could be achieved): activities related to translating the vision into the steps needed to support and favour societal embedding of the PSS concept;
- **actors' network establishment and development** (*WHO* has to be involved in the project): activities related to establishing, managing and developing the network around the project;
- **action plan implementation** (*DO* what planned): activities related to implementing the actions identified in the strategic plan;
- **monitoring, evaluation and learning** (*LEARN* from activities undertaken): activities related to monitoring and evaluating the transition process, and identifying the adjustment to be carried out.

In the design process these five activity clusters run in parallel with the backbone of the process (see Fig. 4.13). Below, a brief description of each activity cluster is provided.

Vision Building and Expectations Shaping The vision consists in a PSS concept developed as an answer to a societal/business challenge. It represents the final goal to be achieved. The vision includes the PSS concept, as well as the motivations to develop it. It should communicate in a clear and effective way what the aims of the project, the main characteristics of the PSS innovation, and its potential (environmental, economic and or socio-ethical) benefits. This cluster comprises activities aimed at building-up a project vision shared between the actors and social groups involved in the project. In particular these activities are related to:

- formalising a PSS concept vision, in order to communicate in a clear and effective way the aims of the project, the main characteristics of the PSS innovation, and its potential (environmental, economic and or socio-ethical) benefits;

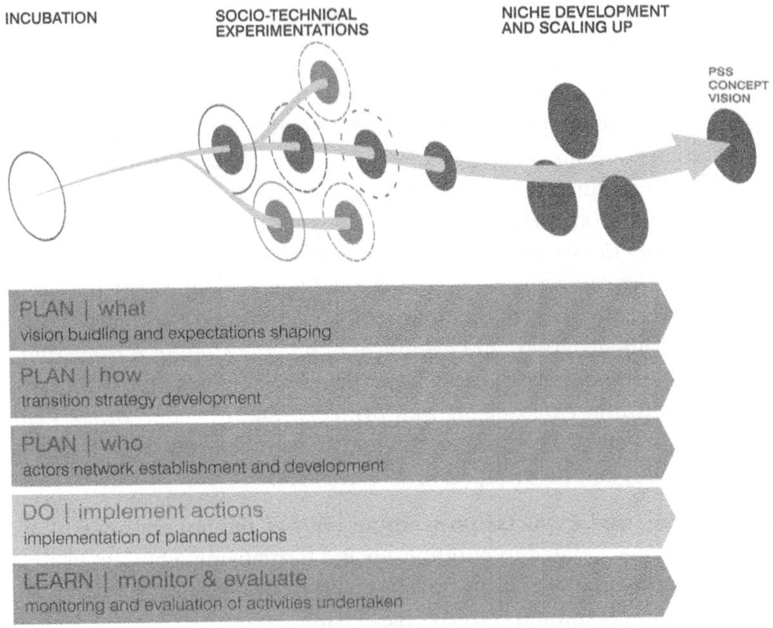

Fig. 4.13 The five activity clusters of the design process. *Source* Ceschin (2012)

- stimulating actors in strategic discussions, in order to encourage them to analyse the vision, identify conflicting issues and potential alternatives;
- stimulating actors to think about the contextual conditions that may favour or hinder the achievement of the vision, and how to strategically influence these conditions;
- continuously adjusting the vision in relation to what is learnt by actors during the different phases of the societal embedding process.

Transition Strategy Development A transition strategy is a hypothesis on how the PSS concept vision could be socially embedded. It is an action plan including the potential steps between the present situation and a future situation in which the PSS concept becomes part of the dominant way in which a societal satisfaction is fulfilled. It translates the long-term goal (the PSS concept vision) into the short and medium-term actions to be undertaken to move towards that goal. It is therefore crucial for coordinating the actions between the different actors involved in the process. In short, a transition strategy provides an action plan which includes: the *actions* to be undertaken and the various *actors* to be involved and the related roles.

As previously underlined the *socio-technical experimentation* phase represents the core of a transition strategy. This phase is preceded by the *incubation* phase, aimed at setting-up the conditions to start the societal embedding process, and can be followed by a *scaling-up phase*, aimed at stabilising the PSS innovation and increasing its influence on the socio-technical regime.

This cluster includes activities related to translating the vision into the steps needed to support and favour the societal embedding of the PSS concept. In particular these activities are related to:

- developing the transition strategy, identifying the milestones to be achieved, the steps to be undertaken, the actors needed to support the path and their roles;
- formalising the transition strategy, in order to effectively visualise and communicate it to the different actors involved in the societal embedding process;
- continuously adjusting the transition strategy in relation to what is learned by actors during the different phases of the societal embedding process.

Actor Network Establishment and Development As deeply illustrated in the previous chapter, the establishment and development of a proper network of actors is recognised as a crucial element in transition processes.

This cluster includes activities focused on the establishment, management and development of a socioeconomic network capable to protect, support and foster radical innovations. In particular these activities are related to:

- selecting the proper actors to be involved during the different phases of the societal embedding process;
- monitoring the actors and social groups directly and indirectly involved in the process and their reciprocal interactions and relations; identifying related convergences, conflicts, interests and power;
- managing the evolution of the network during the societal embedding process.

Implementation of Planned Actions This cluster includes the activities related to the implementation of the socio-technical experiments foreseen in the transition strategy, and the actions required for their subsequent evolution (niche development and scaling-up).

Basically, the process is based on a learning-by-doing philosophy. The continuous implementation of socio-technical experiments (and their development into more stable PSS innovations) represents the crucial step to enable the involved actors to reflect on the activities undertaken and learn how to improve/adjust the PSS innovation and favour its societal embedding.

Monitoring and evaluation of activities undertaken Monitoring and evaluation activities should be focussed on:

- the *socio-technical experiments* and in general the *evolution of the PSS innovation* (technical and usability aspects, acceptability of the PSS by the various social groups involved in the project, implementation and diffusion barriers, etc.);
- the *socio-technical context* in which the PSS innovation is experimented and introduced (opportunities and barriers created by the regime);
- the *network of actors*, their activities and roles, and their changes in way of thinking, doing and behaving;
- the *evolution of the PSS concept vision* and the *transition strategy*.

This cluster includes activities focused on supporting an adequate learning process. In particular these activities are related to:

- *monitoring* the evolution of the PSS innovation and the transition process (actors network, PSS concept vision and transition strategy);
- *reflecting on* and *evaluating* the monitored elements;
- *translating* the evaluation results in *actions to be undertaken*. In fact, through a learning-by-doing approach, evaluation results are used to adjust the project vision (the PSS concept could change in relation to what is learned by actors during the societal embedding process), the action plan, and the actors network (new actors may join and other may leave).

4.4.3 Design Process Overview

On the basis of the structure previously presented, this section puts forward a generic design process to be followed by strategic designers. The design process is illustrated in Fig. 4.14. Below, a short description of the steps included in each phase is provided.

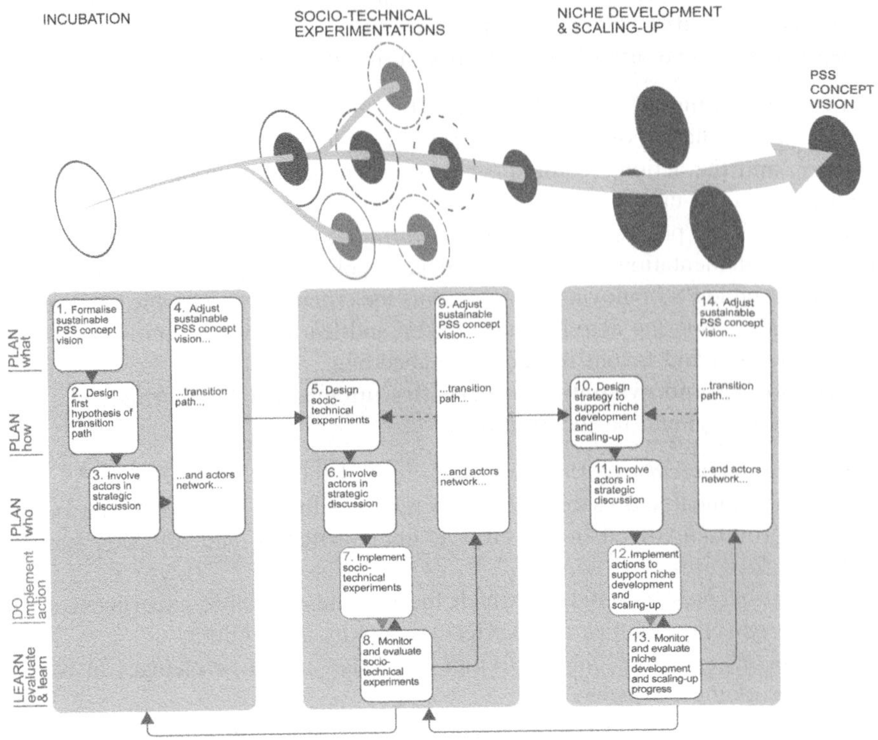

Fig. 4.14 Design process: phases, activity clusters and steps. *Source* Ceschin (2012)

4.4.3.1 Phase 1: Incubation

The incubation phase sets up the conditions to initiate the societal embedding process. The starting point is an eco-efficient PSS concept developed by one or more PSS promoters. The first step is to formalize the project vision in order to communicate in a clear and effective way the aims of the project, the main characteristics of the PSS innovation, and its potential (environmental, economic and/or socio-ethical) benefits.

The next step is to explore how the concept could be socially embedded. This involves making a comparison between the present situation without the PSS and the future situation in which the PSS concept is implemented. It is essential to adopt a strategic attitude in order to identify the short and medium-term actions to be undertaken in order to create the most favourable operable conditions. Actors potentially interested in the concept, as well as actors that could give protection and support to the innovation, are identified. This step produces a first formalized draft action plan (identification of the actions to be undertaken, the actors to be involved, and their roles).

The third step is the involvement of the identified stakeholders in strategic discussions: in this stage meetings and workshops are organized with the actors previously identified. The aim is to collect and confront the different actors' expectations, discuss and adjust both the project vision and the action plan, and achieve a common consensus.

As a result of these meetings and workshops the project vision, the action plan and the actor network are adjusted. The output of this phase is the establishment of a project network that agrees on a project vision and on an action plan.

4.4.3.2 Phase 2: Socio-technical Experimentation

In this phase small-scale socio-technical experiments are implemented in real settings with the aim of learning and exploring how to improve the PSS innovation and how to contribute to its societal embedding.

This phase begins by designing the experiments such that they are protected from the mainstream competition environment and act as Labs, Windows, and Agents of Change (see previous sections). The design of the experiment includes: the identification of an implementation area and the appropriate financial and socio-institutional protections, the technical aspects of the PSS as well as identification of the socio-economic organization to protect and support the innovation (identification of actors to be involved and related tasks).

During the experiment monitoring and evaluation activities are undertaken. These include the experiment (technical and usability aspects, acceptability of the PSS by the various social groups involved in the project, implementation and diffusion barriers, etc.); the actors directly and indirectly involved (their roles, behaviours, expectations, conflicts and convergences); the project vision agreed on and the action plan (and their evolution and adjustment in time).

Evaluation results are then translated into new actions. Through a learning-by-doing approach, the evaluation may lead to an adjustment of the experiment, action plan and the broader project vision, as well as increased understanding of the need for contextual changes and ways to achieve them.

Following this process, different experiments can be executed in different contexts and/or testing new functions.

4.4.3.3 Phase 3: Niche Development and Scaling-up

In this phase the aim is to increase momentum of the PSS innovation (and the related new practices, behaviours and institutions) and begin to influence the socio-technical regime. It is a process that leads the innovation to increasingly become part of the mainstream way in which a societal satisfaction is delivered.

The first step is the identification of appropriate actions to enhance the development and reinforcement of the pathway of experiments. It is essential to link the PSS innovation to a broader context, in order to make it relevant beyond the local scale. Key issues are:

- to repeat the experiments in other contexts and create synergies with similar projects and initiatives. The aim is to share experiences and stimulate and reinforce network building on a broader scale (e.g. within a sector or at a national level);
- to disseminate information/project results and stimulate media attention at a national level;
- to stimulate actors at a strategic level to influence the socio-technical context in order to create the most favourable conditions for the scaling-up of the PSS innovation.

In sum, the aim is to establish deeper linkages with relevant political, industrial and social actors: those that have the power and willingness to directly influence the dominant culture, practices and institutions; those that (in)directly may influence the regime because they have an interest in embedding new sustainable practices; those that can spread information on the PSS innovations; and those that may support the scaling-up of the innovation, such as potential industrial partners, industrial associations or consortia.

During the process monitoring and evaluation activities continuously take place. Evaluation targets include the progress of niche development and scaling-up (e.g. connections with other experiments, enrolment of new actors, dissemination of project results, connections with regime actors, and introduction of the PSS innovation in niche markets), the actors involved, the project vision and the action plan.

The evaluation process can lead to adjustment of the actions to favour niche development and scaling-up, as well as to re-orient the project vision and adjust the actor network and the action plan.

4.4.3.4 Flexibility and Manageability of the Design Process

It has to be underlined that the phases and the steps should not be seen as strictly consecutive stages, but rather as overlapping and interrelated activities. Moreover the process is not meant as a fixed path to be followed, rather as a flexible framework that needs to be adapted in relation to the specificities of each single project. In fact, even if the process proposes a step-by-step support, the actual sequence of steps can be defined project by project and modified during the progress of a project. In other terms the process should be seen as a flexible framework for action which enables strategic designers to guide and orient a societal embedding process by focusing on three main phases and five different clusters of activities.

Also, it must be stressed out that the process presented here should not be considered as a 'recipe for success'. As the *Multi Level Perspective* (*MLP*) warns us, niche development and scaling up require in fact favourable conditions and circumstances (e.g. there should be enough pressure form the landscape, the regime should be sufficiently open to accept radical innovations, etc.). These conditions and circumstances are often outside of the company's direct (or indirect) sphere of influence. Therefore the process from incubation to scaling-up becomes increasingly more uncertain and less manageable, and more influenced by project external events and dynamics.

Projects can only be planned and managed a few years ahead. Outlooks beyond the short term are inherently uncertain. Envisioning, participatory tools and reflexive learning are tools considered useful under such circumstances, but should not be mistaken as to provide blueprints for the future and how to get there.

4.4.4 Supporting Tools

In order to better support practitioners, a set of design tools can be used throughout the design process. These can be grouped in four main clusters, which correspond to four of the five activity clusters (Fig. 4.15). Below, a short description of each tool is provided.[7]

Tools to Formalise PSS Concept Visions These include:

- the *offering diagram*, to shows, through a combination of visual and textual elements, and in a concise form, what the PSS offers to customers (van Halen et al. 2005);
- the *stakeholder system map,* to visualise the socio-economic stakeholders involved in producing and delivering the PSS offer, and their interactions/relations (Jégou et al. 2004);

[7] For a detailed explanation see Ceschin (2012) and the references associated with the tools.

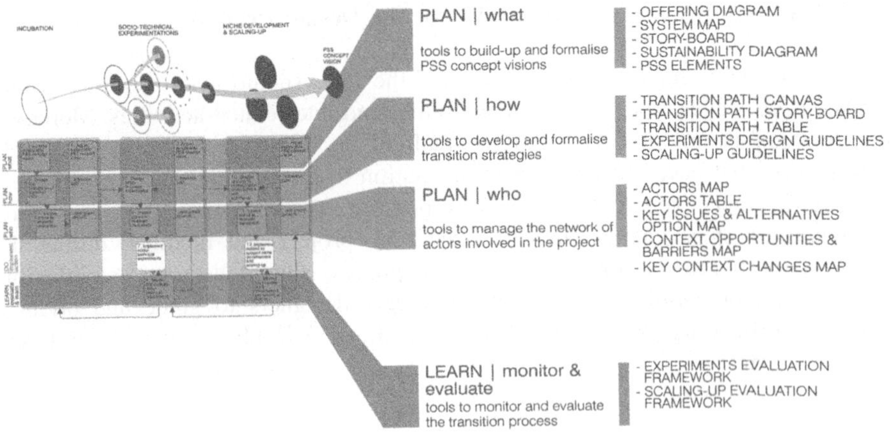

Fig. 4.15 The four clusters of tools. *Source* Ceschin (2012)

- the *storyboard*, to visualise how the PSS offer is delivered to the customers; it chronologically describes the sequence of interactions occurring at front-desk level (interactions between the customer and the offer system) and back-stage level (interactions between the stakeholders involved in producing and delivering the PSS) (van Halen et al. 2005);
- the *PSS elements*, to describe the material and non-material elements (e.g. products, services, communication etc.) required to deliver the PSS offer (Jégou et al. 2004);
- the *sustainability diagram*, to succinctly describe and visualise how the PSS achieves certain sustainability aims (Ceschin and Vezzoli 2007; Vezzoli and Ceschin 2009).

Tools to Design and Formalise Transition Strategies These tools are aimed at supporting the identification of the steps needed to support and favour the societal embedding of the PSS concept. These include (Ceschin 2012):

- the *transition path canvas*, to stimulate the generation of ideas about the action plan to support the introduction and diffusion of the PSS concept;
- the *transition path storyboard*, to visualise the transition strategy in a graphical and succinct form;
- the *transition path table*, to visualise the transition strategy in detailed form;
- the *socio-technical experiments design guidelines*, to orient and support the designing of socio-technical experiments;
- the *scaling up guidelines*, to identify the proper strategies to enhance the development and reinforcement of the pathway of experiments, and favour the scaling-up of the PSS innovation.

Tools to Manage the Network of Actors These include (Ceschin 2012):

- the *actors map*, to visualise the actors and social groups involved in the PSS innovation and the ones that (directly or indirectly) have an influence on it;
- the *actors table*, to make explicit the roles, expectations, convergences and conflicts, interests and power of the actors directly and indirectly involved in the project;
- the *key issues and alternative options map*, to discuss the project vision and action plan with stakeholders and identify conflicting issues and possible alternatives[8];
- the *context opportunities and barriers map*, to stimulate actors to think about the context opportunities and barriers, and how to respectively exploit and overtake them.[9]

Tools to Monitor and Evaluate the Transition Process These tools are aimed at monitoring and evaluating the transition process and identifying the adjustment to be carried out. They include (Ceschin 2012):

- the *experiment evaluation framework*, to stimulate and support the monitoring and evaluation of socio-technical experiments;
- the *scaling up evaluation framework*, to stimulate and support the monitoring and evaluation of scaling-up processes.

4.5 Designing Transition Paths for the Societal Embedding of Sustainable PSSs

In 2003 Manzini and Vezzoli introduced the concept of *strategic design for sustainability*: *"the design of an innovation strategy, shifting the business focus from designing (and selling) physical products only, to designing (and selling) a system of products and services which are jointly capable of fulfilling specific client demands, while re-orienting current unsustainable trends in production and consumption practices"* (Manzini and Vezzoli 2003).

However, as illustrated in this and the previous chapters, there is an urgent need to not only ideate and propose sustainable PSS concepts, but also to favour and support their introduction and scaling up. In respect to this a crucial role can be played by strategic design. Hence, building upon Manzini and Vezzoli's definition, we can describe strategic design for sustainability as *the design of sustainable PSS concepts, and the design and managing of transition strategies to support and hasten their introduction and scaling up* (Ceschin 2012, 2013).

---

[8] Adaptation of the *key issues and alternative table* tool, developed within the ESTEEM project. See Raven et al. (2009).

[9] The tool is an adaptation of the *context opportunities and barriers table* tool, developed within the ESTEEM project. See Raven et al. (2009).

In other terms it is a strategic design capable to guide and support a company, an institution or a network of actors, in the process of ideating sustainable PSS innovations, and in the process of introducing and gradually embedding these innovations in the society. It is an approach in which design, development and implementation should be carried out simultaneously and in continuous interaction. This strategic design approach is based on:

- **envisioning**: the development of long term visions to guide project actors towards the implementation of sustainable radical innovations;
- **experimenting**: the design of sequences of socio-technical experiments to explore and learn how to improve sustainable radical innovations and how to favour their scaling up;
- **involving a wide variety of actors**: the design of the actors network configurations needed to support the societal embedding process, taking in consideration not only the actors belonging to the PSS value chain (producers, partners, suppliers, users, etc.), but also the ones that have the power and willingness to directly and indirectly influence dominant socio-technical regimes;
- **reflexive learning**: the innovation journey is seen as a path based on exploring, searching and learning, capable to bring actors to continuously reflect on and learn from the activities undertaken, and in case reconsider underlying assumptions.

In sum, there is a new important role and challenge for *strategic design for sustainability:* not only to conceive sustainable PSS concepts, but also to understand the contextual conditions in which they are introduced and which strategies and development pathways are the most appropriate. However, the enlargement of the design scope requires designers to be properly equipped. As illustrated in the previous sections designers need:

- a proper **knowledge base**, to be aware of how radical innovations take place and how they can be influenced, oriented and in some ways managed;
- a new **design attitude and skills**, based on the adoption of an experimental-, learning-, and network-based designed and management approach;
- and a practical **design process**, and associated **tools**, to support design practice.

References

Ballon P, Pierson J, Delaere S (2005) Open innovation platforms for broadband services: benchmarking European practices. In: Proceedings of the 16th European Regional Conference, Porto, 4–6 Sept 2005

Borja de Mozota B (1990) Design and management. Les Editions d'Organisation, Paris

Burns C, Cottam H, Vanstone C, Winhall J (2006) RED paper 02: transformation design. Design Council, London

Cantù D (2012) Ideas Sharing Lab. Community centred design for multifunctional and collaborative food services. PhD dissertation, Politecnico di Milano

Ceschin F (2012) The introduction and scaling up of sustainable product-service systems. A new role for strategic design for sustainability. PhD dissertation, Politecnico di Milano

Ceschin F (2013) The societal embedding of sustainable Product-service systems. Looking for synergies between strategic design and transition studies. In: Vezzoli C, Kohtala C, Srinivasan A, Xin L, Fusakul M, Sateesh D, Diehl JC (eds) Product-service system design for sustainability. Greenleaf Publishing, Sheffield (in press)

Ceschin F, Vezzoli C (2007) Sustainable product-service systems development. Application and evaluation of the SDOS (Sustainability Design-orienting Scenario) methodology for the upstream waste reduction in the food and in the paper chains in the city of Brescia. In: Proceedings of the 11th European roundtable on sustainable consumption and production (ERSCP 2007), FHNW (School of Life Sciences—Institute for Ecopreneurship), Basel, 20–22 June 2007

Gaziulusoy Aİ (2010) System innovation for sustainability: a scenario method and a workshop process for product development teams. PhD thesis, The University of Auckland

Gaziulusoy Aİ, Boyle C, McDowall R (2013) System innovation for sustainability: a systemic double-flow scenario method for companies. J Cleaner Prod 45:104–116

Gumus H (2009) Kanga: a sustainable system design for the transportation of learners with disabilities in Cape Town—South Africa. Master's degree thesis, Politecnico di Milano, Faculty of Design

Hegger DLT, van Vliet J, van Vliet BJM (2007) Niche Management and its Contribution to regime change: the case of innovation in sanitation. Technol Anal Strateg Manage 19:729–746

Hillgren PA, Seravalli A, Emilson A (2011) Prototyping and infrastructuring in design of social innovation. CoDesign: Int J CoCreation Des Arts 7:169–183

Hoogma R (2000) Exploiting technological niches: strategies for experimental introduction of electric vehicles. PhD Thesis. Twente University Press, Enschede

Hoogma R, Kemp R, Schot J, Truffer B (2002) Experimenting for sustainable transport: the approach of strategic niche management. Spon Press, London

Jégou F (2011) Social innovations and regional acupuncture towards sustainability. Chin J Des 214:56–61

Jégou F, Manzini E (eds) (2008) Collaborative services. Social innovation and design for sustainability. Edizioni Polidesign, Milan

Jégou F, Manzini E, Meroni A (2004) Design plan. A design tool-box to facilitate solution oriented partnerships'. In: Manzini E, Evans S, Collina L (eds) Solution oriented partnership. Cranfield University, Cranfield, pp 107–118

Kemp R, Loorbach D (2003) Governance for sustainability through transition management. In: Proceedings of the international human dimensions programme (IHDP) conference, Montreal, Oct 2003

Kemp R, Schot J, Hoogma R (1998) Regime shifts to sustainability through processes of niche formation: the approach of strategic niche management. Technol Anal Strateg Manage 10:175–195

Kemp R, Rip A, Schot J (2001) Constructing transition paths through the management of niches. In: Garud R, Karnøe P (eds) Path dependence and creation. Lawrence Erlbaum, London

Loorbach D (2002) Transition management: governance for sustainability. In: Proceedings of the international dimensions of human change conference, Berlin, 2002

Loorbach D (2007) Transition management: new mode of governance for sustainable development. PhD dissertation, Erasmus University Rotterdam

Loorbach D, Rotmans J (2006) Managing transitions for sustainable development. In: Olsthoorn X, Wieckzorek A (eds) Understanding industrial transformation: views from different disciplines. Springer, Dordrecht, pp 187–206

Manzini E (1999) Sustainable solutions 2020—Systems. In: Proceedings of the 4th international conference towards sustainable product design, Borschette Conference Centre, Brussels, 12–13 July 1999

Manzini E, Rizzo F (2011) Small projects/large changes: participatory design as an open participated process. CoDesign: Int J CoCreation Des Arts 7:169–183
Manzini E, Vezzoli C (2003) A strategic design approach to develop sustainable product service systems: examples taken from the 'environmentally friendly innovation' Italian prize. J Cleaner Prod 11:851–857
Mellick Lopes A, Fam D, Williams J (2011) Designing sustainable sanitation: involving design in innovative, transdisciplinary research. Des Stud 33:298–317
Meroni A (ed) (2007) Creative communities. People inventing sustainable ways of living. Edizioni Polidesign, Milan
Meroni A (2008a) Strategic design: where are we now? Reflection around the foundations of a recent discipline. Strateg Des Res J 1:31–38
Meroni A (2008b) Strategic design to take care of the territory. Networking creative communities to link people and places in a scenario of sustainable development. Keynote presented at the P&D Design 2008—8° Congresso Brasileiro de Pesquisa e Desenvolvimento em Design, Campus Santo Amaro, San Paolo
Mulgan J (2009) Strengths, weaknesses and a way forward? Available online: www.socialinnovationexchange.org/designforsi/blog?page=1. Accessed 21 Feb 2013
Raven RPJM (2005) Strategic niche management for biomass. PhD dissertation, Technische Universiteit Eindhoven, Eindhoven
Raven RPJM, Jolivet E, Mourik RM, Feenstra YCFJ (2009) ESTEEM: Managing societal acceptance in new energy projects. A toolbox method for project managers. Technol Forecast Soc Chang 76:963–977
Rotmans J, Kemp R, van Asselt M, Geels FW, Verbong G, Molendijk K (2000) Transities & transitiemanagement: De casus van een emissiearme energievoorziening (Transitions & transition management: The casus of a low emission energy supply). Maastricht University, Maastricht
Sangiorgi D (2010) Transformative services and transformation design. Int J Des 5:29–40
Simeone G, Cantú D (2011) Feeding Milan. Energies for change. A framework project for sustainable regional development based on for de-mediation and multifunctionality as design strategies. Cumulus Shanghai Conference 2010: Young Creators for Better City and Better Life, Shanghai, 7–10 Sep 2011
Stålbröst A (2008) Forming future IT: the living lab way of user involvement. Doctoral thesis, Luleå University of Technology
Van de Poel I (2000) On the role of outsiders in technical development. Technol Anal Strateg Manage 12:383–397
Van Halen C, Vezzoli C, Wimmer R (2005) Methodology for product service system. How to develop clean, clever and competitive strategies in companies. Van Gorcum, Assen
Vezzoli C (2007) System design for sustainability. Theory, methods and tools for a sustainable "satisfaction-system" design. Maggioli Editore, Rimini
Vezzoli C, Ceschin F (2009) Sustainable scenarios for upstream waste reduction in the food chain. In: Hülsen I, Ohnesorge E (eds) Food science research and technology. Nova Publishers, New York
Vezzoli C, Ceschin F, Kemp R (2008) Designing transition paths for the diffusion of sustainable system innovations. A new potential role for design in transition management? In: Cipolla C, Peruccio PP (eds) Changing the change. Design, visions, proposals and tools. Changing the change conference, Turin, 10–12 Jul 2008. Umberto Allemandi, Turin
Weber M, Hoogma R, Lane B, Schot J (1999) Experimenting with sustainable transport innovations: a workbook for strategic niche management. Universiteit Twente, Twente
Zurlo F (1999) Un modello di lettura per il Design Strategico. La relazione tra design e strategia nell'impresa contemporanea. PhD thesis, Politecnico di Milano

Chapter 5
Where Do We Go from Here?

Abstract This chapter summarises the main findings and lessons learned, highlights and generalises the main contributions, and indicates potential paths for future research.

Keywords Product-service system · Sustainable development · Transition studies · Strategic design · Design for sustainability · Strategic niche management · Transition management · Research challenges · Research agenda

5.1 Research Findings

Sustainable PSS innovation represents a valuable concept for enhancing company competitiveness and at the same time providing environmental benefits. However, these innovations are in most cases radical, and their introduction and diffusion usually encounter the opposition of existing customers' habits (cultural barriers), companies' organizational structures (corporate barriers), and regulative frameworks (regulative barriers). Hence, if immediately exposed to the mainstream market environment, it is highly probable they will not survive. An important challenge is therefore not only to conceive sustainable PSS concepts, but also to understand the contextual conditions in which they are introduced and explore the most suitable strategies and development pathways to embed these concepts in society.

This raises important questions on the role of design in addressing this challenge. And we can observe that this is becoming a more and more relevant issue in the Design for Sustainability community. In this respect researchers have started to explore the role of design in relation to large-scale changes. Manzini, Meroni and Jégou have proposed three broad design strategies to scale-up social innovations: *enabling*, *replicating* and *synergising* (Jégou and Manzini 2008; Meroni 2008; Jégou 2011). Sangiorgi (2010) has explored the role of service design as an engine for wider societal transformations. Mellick Lopes et al. (2011) have investigated the potential of visual communication design to facilitate social learning in the

F. Ceschin, *Sustainable Product-Service Systems*, PoliMI SpringerBriefs,
DOI: 10.1007/978-3-319-03795-0_5,

transition to more sustainable systems. Joore (2010) developed a Multilevel Design Model to clarify the mutual relationship between new products, PSSs, socio-technical systems and societal changes. Gaziulusoy (2010) and Gaziulusoy et al. (2013) have developed a scenario method to link activities/decisions at the product development level in companies with the transformation which needs to take place at the societal level to achieve sustainability.

The original contribution of this book is to combine design thinking with concepts and insights from transition studies.[1] Building upon insights from the *transition studies* field, this book put forward a conceptual framework for the introduction and scaling-up of sustainable PSSs. A crucial role is given to the implementation of sequences of socio-technical experiments, partially protected spaces where broad networks of actors incubate, test, develop and bring the innovation to maturity without the direct pressure coming from the market environment. Theoretical and empirical evidence supports the proposal that, in order to effectively contribute to transition processes, socio-technical experiments should be conceived as *Labs*, *Windows* and *Agents of Change*. Strategic design could thereby play a role not only in generating eco-efficient PSS concepts, but also in designing transition paths to support and facilitate the introduction and scaling-up of the concept itself. The book also discussed the new design approach and new design capabilities required by designers to operate at such a level.

In short, the core contributions of this research are:

- First, a **conceptual framework** describing and explaining how sustainable PSS innovations take place was developed. It reduces the complexity of the process and provides a comprehensive overview of the main influencing factors. The framework describes the process of introduction and scaling-up as an evolutionary transition path, characterised by an incubation phase, a socio-technical experimentation phase and a scaling-up phase. It also brings all the influencing factors together, grouping them in four main clusters (factors related to the societal embedding process, the actor network, the project vision and actor's expectations, and the learning process).
- Second, the research contributed to clarify how **socio-technical experiments** can trigger the process of introduction and diffusion of sustainable PSS innovations. Building upon insight from transition studies and the result of the first part of the research, a definition of socio-technical experiment was developed. It identifies the key issues to be addressed when designing and managing this kind of experiments.
- Third, the research conceptualised the *functions* that socio-technical experiments can play in transition processes: experiments can act as **Labs** (to test, learn about and improve the PSS innovation on multiple dimensions), **Windows** (to raise interest in the innovation project and the related actors, disseminate results, build up synergies with existing similar projects/initiatives, and attract

[1] This topic was initially explored in Vezzoli et al. (2008) and further developed in Ceschin (2010, 2012) and Ceschin et al. (2011).

and enrol new actors) and **Agents of change** (to influence contextual conditions in order to promote and quicken the transitioning process). This conceptualisation can be used as a guide for identifying which elements need to be designed and developed in socio-technical experiments.
- Fourth, the research proposed a **new role for strategic design for sustainability**: a role that couples the generation of sustainable PSS concepts with the design and manage of transition strategies to support and speed up their introduction and scaling up.
- Fifth, on the basis of this new role, the research contributed to clarify the new **design attitude and capabilities** needed by strategic designers to operate at such strategic level. It also identified the design knowledge base that strategic designers need to be equipped with.
- Finally, on a more operational point of view, the research developed a practical "how to do it" **design process**, and associated guidelines and tools, to support strategic designers, project managers, and consultants in designing and managing the societal embedding process of sustainable PSS innovations.

5.2 Research Limitations

The first limitation concerns the limited application of the developed design approach and associated tools. The design approach and tools were in fact applied only in one project, which is of course a small sample for final conclusions. This limitation is inherently due to the complexity and the timescale related to the process of introduction and diffusion of sustainable PSSs: in fact it might take several years to follow the implementation and scaling up of socio-technical experiments (e.g. the Cape Town Sustainable Mobility project was launched in 2009 and it is still running). Thus, given the limited extent of time of this Ph.D. research (3 years) it was not possible to entirely test the design approach and tools in a *real time perspective*. However, this limitation was foreseen in the beginning of the research and thus the research methodology was properly designed in order to deal with it. In fact, practitioners and academic experts were involved to assess the design approach and tools in terms of their potential *practicality* (approach and tools are practically usable in the settings for which they have been designed and developed) and *effectiveness* (the use of the approach and tools brings to desired outcomes). Anyhow, it is of course clear that more research would be needed to further test the design approach and tools on practice.

A second limitation is related to the researcher's role. In both the first and the second part of the research the author was deeply involved with the investigated phenomenon. During the case study analysis the researcher directly interacted and collected data from persons involved in the cases. During the second stage of the research the researcher was directly involved (by close collaboration with other designers and practitioners) in the development, testing and refinement of the design approach and tools. The fact that the researcher was directly involved in the studied phenomenon might be seen as a limitation. A deep involvement could in fact hinder researcher's independence and objectivity. In order to limit this risk, several external stakeholders (practitioners and academic experts in the field of PSS and transition studies) were involved to assess and validate the results of the research. These stakeholders provided an independent evaluation, not influenced by the opinions and view of the researcher. In addition it must be stressed that this deep involvement had some positive effects: the advantage was in fact that the researcher had the opportunity to access rich data and go closer to practice.

The third limitation of the research is related to the case studies analysed in the first part of the research. These cases are related to the innovation journeys made by six European companies in introducing their sustainable PSS innovations in the market, and they are characterised by heterogeneity. In fact, the PSS innovations analysed can be classified under the two different typologies: *user oriented PSS* and *result oriented PSS*. Moreover these PSSs differ in terms of business sector and companies size. This heterogeneity might be seen as a limitation. In fact it could be argued that it would have been more valuable to select homogeneous cases with similar characteristics (e.g. cases related to a given business sector and/or PSS typology). Of course this choice would have contributed to enhance internal validity but, since this study has an explorative aim (little attention has been put by literature on this issue), it was decided to select cases in order to have a broader picture and facilitate the generalisability of results.

5.3 Future Research Directions

5.3.1 Improving and Refining the Results Achieved in this Research

As previously illustrated, the design approach and tools have been so far applied only in a limited number of projects. Thus, an important issue is to make them more reliable. In respect to this, two main recommendations can be made:

The first recommendation is to further apply and test the design approach and tools in other projects related to the societal embedding of sustainable PSSs. In particular it would be valuable to apply them in different situations:

- in different market sectors (e.g. PSSs in the energy, food, mobility sectors);
- and with different actors as main promoters of the PSS (e.g. project in which the main promoter is a multinational company, a small/medium enterprise, an NGO, a public institution, a research/educational institution, etc.).

It is also suggested to track these projects for several years in order to monitor their impacts (i.e. diffusion of the PSS solution). This combination of testing in different situations (and with a long time perspective) could lead to improve and refine the design approach and tools and also to produce insights on how to adapt them in relation to the needs and characteristics of specific target groups (e.g. for a multinational company working in the energy sector, or for a small company working in the mobility sector).

The second recommendation is about the historical cases related to the introduction and scaling up of sustainable PSSs. In this research the case studies were characterised by heterogeneity (different PSS typologies, business sectors and companies sizes) and therefore they allowed the development of a general conceptual framework. It would be interesting to enlarge the number of cases and explore the differences of the societal embedding process in relation to the type of PSS, the business sector, etc. This could help create a repository of historical cases and potentially identify patterns of introduction and diffusion in relation to different conditions (e.g. in relation to specific business sectors and/or PSS type). Moreover this kind of results could be used to improve and refine the design approach and tools and adapt them in relation to specific conditions. For example specific guidelines and tools, to be included in the general design approach, could be developed in relation to specific business sectors and/or PSS type, etc.

5.3.2 Applying the Research Results in Other Fields

It is expected that the findings of this research (in particular the design approach and tools) could be generalised and used not only in relation to sustainable PSSs but also to other kinds of radical innovations. An example is represented by sustainable social innovations. Social innovations present interesting potentialities in terms of new and more sustainable ways of living. However the crucial issue is to stimulate a wider adoption of these solutions (Manzini 2011). Like sustainable PSSs, sustainable social innovations are solutions that change (in same cases even radically) the traditional way of fulfilling needs, affecting the spheres of production, distribution and consumption. Therefore the process to scale up these innovations challenges current thinking, behaviour, routines, regulations, etc. For this reason, the insights produced in this research might be useful to stimulate the introduction and diffusion of this kind of innovation. For example the *socio-technical experiment* concept as well as the design approach and tools could be adapted and used to spread and scale up this kind of solutions. However, it must be underlined that there are some important differences between PSSs and social innovations, and this could limit the application and adaptation of the research results. For example PSSs innovations are often initiated and proposed by a company (or by a small network of actors including one or more company), while social innovations emerge from a variety of actors directly involved in conceiving and implementing the solution. In social innovations final users and communities play a key role in developing and self-managing the solutions. Thus, it is clear that the role of the designer in supporting the introduction and scaling up process could be different in the two cases. In particular the way of approaching and interacting with the actors involved in project would probably change.

References

Ceschin F (2010) How to facilitate the implementation and diffusion of sustainable product-service systems? Looking for synergies between strategic design and innovation sciences. In: Ceschin F, Vezzoli C, Zhang J (eds) Sustainability in design: now! Challenges and opportunities for design research, education and practice in the XXI century. Proceedings of the learning network on sustainability (LeNS) conference, Bangalore, 29 Sep–1 Oct 2010, vol 1. Greenleaf Publishing, Sheffield, pp 440–454

Ceschin F (2012) The introduction and scaling up of sustainable product-service systems. A new role for strategic design for sustainability. Ph.D. Dissertation, Politecnico di Milano

Ceschin F, Vezzoli C, Vergragt P (2011) Small scale socio-technical experiments as stepping stones for eco-efficient product-service systems diffusion: a new role for strategic design for sustainability. In: Hesselbach J, Herrmann C (eds) Functional thinking for value creation. Proceedings of the 3rd CIRP international conference on industrial product service systems. Technische Universität Braunschweig, Braunschweig, pp 5–6 May 2011

Gaziulusoy Aİ (2010) System innovation for sustainability: a scenario method and a workshop process for product development teams. Ph.D. Thesis, The University of Auckland

Gaziulusoy Aİ, Boyle C, McDowall R (2013) System innovation for sustainability: a systemic double-flow scenario method for companies. J Cleaner Prod 45:104–116

Jégou F (2011) Social innovations and regional acupuncture towards sustainability. Chin J Des 214:56–61

Jégou F, Manzini E (eds) (2008) Collaborative services. Social innovation and design for sustainability. Edizioni Polidesign, Milan

Joore P (2010) New to improve: The mutual influence between new products and societal change processes. Ph.D. Thesis, Delft University of Technology

Manzini E (2011) To make things happen. Key-note speech at the design pleasurable product interfaces (DPPI11) international conference, Politecnico di Milano, Milan, 22–25 Jun 2011

Mellick Lopes A, Fam D, Williams J (2011) Designing sustainable sanitation: involving design in innovative, transdisciplinary research. Des Stud 33:298–317

Meroni A (2008) Strategic design to take care of the territory. Networking creative communities to link people and places in a scenario of sustainable development. Keynote presented at the P&D design 2008—8° Congresso Brasileiro de Pesquisa e Desenvolvimento em design, Campus Santo Amaro, San Paolo

Sangiorgi D (2010) Transformative services and transformation design. Int J Des 5:29–40

Vezzoli C, Ceschin F, Kemp R (2008) Designing transition paths for the diffusion of sustainable system innovations. A new potential role for design in transition management? In: Cipolla C, Peruccio PP (eds) Changing the change. Design, visions, proposals and tools. Changing the change conference, Turin. Umberto Allemandi, Turin, 10–12 July 2008